SHEET METAL PRACTICE
SECOND EDITION si metric

SHEET METAL PRACTICE
SECOND EDITION si metric

BILL NEUNDORF
Master Teacher
York County Board of Education
Aurora, Ontario

CLAUDE STEVENS, B.A.
Sheet Metal Instructor

THIS TEXT CONFORMS TO THE SI SYSTEM OF MEASUREMENT

With technical illustrations by:

ALAN McDERMOTT
Technical Director
Thornhill Secondary School
Thornhill, Ontario

and

DON McCLENNAN
Assistant Technical Director
Eastdale Secondary School
Oshawa, Ontario

McGraw-Hill Ryerson Limited

Toronto, Montreal, New York, St. Louis, San Francisco, Auckland, Bogotá,
Düsseldorf, Johannesburg, London, Madrid, Mexico, New Delhi, Panama,
Paris, São Paulo, Singapore, Sydney, Tokyo

SHEET METAL PRACTICE, Second Edition

Copyright © McGraw-Hill Ryerson Limited, 1977
Copyright © McGraw-Hill Company of Canada Limited, 1962
All rights reserved. No part of this publication may be reproduced, stored in a retrieval system, or transmitted, in any form, or by any means, mechanical, electronic, photocopying, recording or otherwise, without the prior written permission of McGraw-Hill Ryerson Limited.

234567890 MP 54321098

Printed and bound in Canada

Canadian Cataloguing in Publication Data

Neundorf, William, 1925-
  Sheet metal practice

ISBN 0-07-077681-4

1. Sheet-metal work. I. Stevens, Claude, 1931-
II. Title.

TS250.N49 1977   671.8'23   C77-001103-9   NOV 13 '78

Table of Contents

Preface ... vii

INTRODUCTION TO SHEET METAL ... viii

SHEET METAL PRACTICE AND METRIC CONVERSION ... viii

PART ONE — MATERIALS, TOOLS AND OPERATIONS ... 1

1. SHOP SAFETY RULES ... 1

2. COMMON SHEET METALS ... 2
- BASE METALS
- COATED METALS
- ALLOYS
- PROPERTIES OF METALS

3. HAND TOOLS ... 4

4. BENCH STAKES ... 6

5. PROFILE VIEWS ... 8

6. HAND OPERATIONS ... 12
- SCRIBING A LINE ... 12
- STRAIGHT CUTTING AND NOTCHING ... 12
- SOLDERING ... 14
  - Solder
  - Soldering Fluxes
  - Soldering Irons
  - Gas Furnaces
  - Soldering a Lap Joint
- HAND GROOVING ... 18
  - Grooved Seam Allowance
  - Finishing the Grooved Seam
- WIRING ... 20
  - The Wired Edge
  - Wiring a Straight Edge
- RIVETING ... 21
  - Types of Hand Rivets
  - Rivet Sets
  - Steps in Riveting
  - Blind Riveting

PUNCHING ... 25
- Hand Lever Punch
- Principal Parts
- Changing the Punch and Die
- Punching Holes

7. MACHINES AND MACHINE OPERATION ... 27

- THE FOOT OPERATED SQUARING SHEARS ... 27
  - Principal Parts
  - Securing a Working Edge
- THE BAR FOLDER ... 28
  - Principal Parts
  - Forming a Single Fold
  - Types of Folds
- THE STANDARD HAND BRAKE ... 32
  - Principal Parts
  - Forming a Right Angle Bend
- THE BOX AND PAN BRAKE ... 34
  - Principal Parts
  - Forming a Box
- THE SLIP-ROLL FORMING MACHINE ... 36
  - Principal Parts
  - Rolling Cylindrical Shapes
- THE DRILL PRESS ... 38
  - Principal Parts
  - Drilling Small Holes
- THE BAR AND TUBE BENDER ... 40
  - Principal Parts
  - Scroll Bending
  - Square Bending
- THE ROD PARTER ... 42
  - Principal Parts
  - Gauge Operation
- THE LOCKFORMER ... 43
  - Principal Parts
  - Maintenance
  - Forming a Pittsburgh Lock
  - Operating the Power Flanging Attachment
- THE TURNING MACHINE ... 45
  - Principal Parts
  - Turning an Edge for Wiring

THE BURRING MACHINE	48	FIXED HANDLE DUSTPAN	77
Principal Parts		CAMP CUP	80
Burring a Disc			
THE EASY EDGER	50	HANGING PLANTER	82
Principal Parts		WINDOWSILL PLANTER	84
Turning a Right Angle Edge			
THE TURRET PUNCH PRESS	52	CLOTHES PEG BOX	86
Principal Parts			
Maintenance		CORNER SHELF	90
Punching Procedure			
THE SPOT WELDER	53	THREE RING BOOK BINDER	92
Principal Parts		COAL SCUTTLE ASH TRAY	94
Operating Procedure			
		AQUARIUM	96
PART TWO — PROJECTS	57	FISHING TACKLE BOX	98
RECTANGULAR BOX 1	58	WATERING CAN	101
RECTANGULAR BOX 2	60	SHOESHINE BOX	104
COOKIE SHEET	61	FILING CABINET	107
SPIKE FILE	63	FILING TRAY	110
SPICE RACK	65	CARRY ALL	113
UTILITY BOX	67	TOOL BOX	115
WASTE BASKET 1	70	BARBECUE	118
WASTE BASKET 2	72	PICNIC COOLER	124
LEVERED HANDLE DUSTPAN	74	FUNNEL	131

PREFACE

Since the introduction of automation to the craft trades, many hand skills have disappeared into history. However, the fundamentals of both hand and machine operations are required for a thorough knowledge of any craft. With this thought in mind, the beginner will find this book a rewarding introduction to sheet metal practice.

Subject matter is arranged in two main parts. The beginner will first learn how to perform various hand and machine operations; he then has the opportunity to apply this information to a series of useful projects. To aid him in review, related assignment questions follow the text material.

All operations are arranged progressively to the degree of difficulty that will be encountered in the layout and assembly of the projects. Illustrations aid in the development of visualizing powers as the beginner executes the tasks assigned by his instructor.

The chapters on materials and hand and machine operations will enlighten students and prove useful in helping them to design some of their own projects. In order to achieve this objective, students should learn the basic skills well.

Once these skills are mastered students may feel free to expand upon the projects given and create new ones. Creativity and a sense of pride in workmanship are two of the rewards of well-developed skills.

We, the authors wish to acknowledge the co-operation of the New York Bureau of Industrial and Technical Education; the New York State Education Department; the Brown Boggs Foundry and Machine Company, Limited; Dreis and Krump Manufacturing Company; the Marathon Equipment and Supply Company; Diacro Division, Houdaille Industries, Inc.; United Shoe Machinery Co.; Dominion Foundries and Steel Ltd. (DOFASCO); F. H. Welding Co.; Mr. Ronald Scovell, Bathurst Heights Collegiate and Vocational School, for drawings of hand tools, the box and pan brake; Mr. Cyril F. Marsh, deceased, former sheet metal instructor at Danforth Technical School, for information and guidance, and Mr. Tom Shields, technical education editor, McGraw-Hill Ryerson Ltd., for overseeing the metrication of this edition and providing the section on Sheet Metal Practice and Metric Conversion.

INTRODUCTION TO SHEET METAL

In any trade, the craftsman first learns the basic skills of both hand and machine operations and the advantages and limitations of various materials. Then he is able to adapt this knowledge to construct a specific project. When a sheet metal mechanic starts a job, he uses the facts and skills he has learned to solve the problems involved in that job.

This course incorporates the fundamentals of the sheet metal trade. A thorough study of the course requires hours of concentrated thought and practical work. Effort on your part in the fabrication of projects will show if you are adapted to this trade. It is hoped that you will develop sound reasoning, clear thinking, good judgment, and powers of concentration, and that you will also develop professional work habits — neatness, accuracy, initiative, perseverance, and self-reliance.

Do not become discouraged; this course has been established to test you, not to condemn you. Concentrate on learning the reasons behind each step of every operation and you will master the fundamentals of sheet metal.

SHEET METAL PRACTICE AND METRIC CONVERSION

THE NEED FOR METRIC CONVERSION

North America is the only area of the world today which has not fully converted to a metric system of measurement. *Inch-pound* systems of measurement, (such as Imperial, Canadian and U.S. Customary) have been used almost exclusively instead. Two important facts, however, are making conversion to a metric system of measurement a current reality. First, trade between North American, European, and Asian countries has increased dramatically since World War II, necessitating a universal measurement system and world standards for manufacture of many parts and goods. Second, the *inch-pound* systems of measurement are made up of poorly related units with awkward subdivisions and fractions, making the system difficult to master and difficult to perform calculations rapidly and accurately.

The early metric systems were much easier to learn and use in comparison to the *inch-pound* systems, but they too had disadvantages, largely resulting from the fact that each metric country had its own special system. The systems differed in how they defined their units, in the number and type of units used, and in other matters. These differences made it difficult for a draftsman in Germany to understand a drawing from Japan, for example.

The metric nations of the world wanted a universal metric system. In 1960 the International System of Units, more commonly known as SI from the French Système International d'Unités, was created to satisfy this need. Superfluous units were discarded, the remaining units were carefully defined, and a special international style of writing SI was developed. In the years from 1960 to the present metric nations have been converting to SI and the North American countries, encouraged by the benefits of having one world-wide system, have passed laws to adopt SI for all facets of life. Many large industries are already converting their drawings and equipment. Manufacturers of consumer products, such as sugar and shoes, are also engaged in conversion. Today's sheet metal student, therefore, must expect to live and work in an SI metric world.

In SI quantity units exist for different types of measurement. Examples are the metre (m) for length, the kilogram (kg) for mass, and the Kelvin (K) for temperature. In everyday use the degree Celsius (°C) is used instead of the Kelvin. These units, with the exception of the Kelvin and degree Celsius, may be multiplied or divided by factors of ten to make larger or smaller units. The factors have different prefix names, which are applied to the name for the base unit. For example, the SI base unit for length, the metre, is just a bit higher than the hip joint on a man of average height. In order to work in length units larger or smaller than the metre we would have to employ one of the following prefixes:

Name	Meaning	Symbol
mega-	1 000 000	M
kilo-	1 000	k
hecto-	100	h
deca-	10	da
deci-	0.1	d
centi-	0.01	c
milli-	0.001	m
micro-	0.000 001	μ

In other words, we could talk of kilometres, millimetres, or micrometres if we found it useful. A kilometre would be equal to 1000 metres, whereas a millimetre would only measure one-thousandth of a metre. Ten millimetres and one centimetre would measure the same length.

METAL PRODUCTS AND SHEET METAL PRACTICE

Presently all or most of the materials in the school shop are made to *inch-pound* standards and therefore come in *inch-pound* sizes. The metal industries which produce these products are only now beginning the switch to SI standards and sizes. Because of the expense involved for these industries it may be some time before the new sizes are available for use in schools. Two new standards from the Canadian Standards Association, *CSA 312.1-Preferred dimensions of flat metal products,* and *CSA 312.2-Preferred dimensions for round, square, rectangular, and hexagonal metal products* should be available by the time this book is published. These documents will indicate what the new sizes will be, but not when they will be produced. It is likely that your school will have to work with materials and tools manufactured to *inch-pound* sizes for several years yet.

It is important not to confuse metal products manufactured to *inch-pound* sizes with products manufactured to SI sizes. It is also important not to call the products manufactured to one measurement system by the equivalent units of the other system. A ¼ in. welding rod is not a 6 mm or even a 6.35 mm rod — to call it thus would lead people to believe that this was the metric standard size, which it may not be. Always call *inch-pound* products by their *inch-pound* names, and metric products by their metric names.

Wherever possible, SI units have been used exclusively in this text. All drawings are dimensioned in millimetres and incorporate the new symbols for diameter (ø) and radius (R) where applicable. *Inch-pound* units are used only in reference to *inch-pound* products or tools which are presently bought, sold, or manufactured only to *inch-pound* standards. The sheet metal student must therefore think and work in SI units while sometimes using products and tools built to *inch-pound* standards. Although proper SI style has been used throughout the text for most applications, there is one exception. *Inch-pound* decimal quantities of a value of less than 1.00 (e.g. .03 in.) are given *without* a zero in front of the decimal point. All other decimal quantities or dimensions are given with the zero in front of the decimal point (e.g. 0.43 mm).

By separating the use of the two systems in this manner students will clearly see their differences. Confusion will be minimal, while the problems conversion creates for the community and individual as well as the potential advantages of several aspects of SI will be easily appreciated.

Conversion Tables

It should rarely be necessary for the sheet metal student to convert from one measurement system to another. For this reason, no conversion tables are given here. The student and his teacher are both referred to the *Canadian Standards Association Metric Practice Guide-CSA Z234.1* for precise conversion factors.

SHEET METAL PRACTICE AND THE SI SYSTEM

Students will normally use only a few types of measurement in sheet metal work. The quantities we usually want to measure are length, area, temperature, and mass. Other quantities, such as force and pressure, are sometimes encountered as well.

Length

Although it would be possible to use all the prefixes listed above in combination with the metre, not all of them would prove very useful in sheet metal work. It would be easiest to restrict ourselves to the really essential ones. For this reason, sheet metal producers and people employed in sheet metal work will be talking, when they refer to length of metal stock, in either millimetres (mm) or metres (m). A millimetre is about the width of the side of a dime. An example has already been given for the metre.

Area

Area is a surface enclosed by a line or group of lines. Circles and squares are areas. Area is measured using squared units of length.

Square millimetres (mm^2), square centimetres (cm^2), and square metres (m^2) are the most common units and prefixes. The fingernail on your index finger probably has an area of about 1 cm^2. The erasers on pencils have an end surface area of about 30 mm^2. A square metre is about the area of half of one side of a door of average height.

Temperature

Temperature is measured in degrees Celsius (°C). 0°C is the freezing point of water, whereas water boils at 100°C. 20°C is about room temperature, 10°C is cool but temperate, and 30°C is a warm summer day. −10°C is fine winter weather, −20°C is about the point where we feel the sharpness in the air, and −30°C is biting cold, cold enough to make us feel that it's hard to breathe. No prefixes are used with temperature measurement. Metals which have to be heated may reach temperatures of 200°C or much greater extremes, depending on the process involved.

Mass

Mass refers to the amount of material or matter in an object. Small masses are usually measured in either kilograms (kg) or grams (g). The megagram (Mg) is widely called a metric tonne (t). Many products are now packaged by the kilogram. It must be emphasized that a kilogram of lead on earth will still have a mass of 1 kg in space. The lead in space, however, will not have weight, as there is no gravity in space. This illustrates the difference between mass and weight. The word "weight" refers only to the force of gravity acting on a mass. If something helps to counteract the force of gravity on a mass it will not weigh as much. Masses suspended on springs, such as a car body, are an example. Almost everyone presently uses the term "weight" incorrectly to mean "mass". Try to use the term "mass" wherever possible, as it is usually correct. Study the following examples:

The springs help support the "weight" of the car.
The "paperweight" on her desk has a mass of 200 g.
The "heavyweight" boxer had a mass of 100 kg.

Try making up your own examples and ask your teacher to check them.

Volume or Capacity

The capacity of some object such as a box, bottle, pail, etc. to hold some other material like water or earth is called volume. Volume can also refer to the space enclosed by the walls of some object. The cubic decimetre (dm^3) is the SI standard unit of volume. This unit is commonly called a litre (ℓ). One thousand millilitres (ml^3) are contained in a litre. The litre and millilitre are the two most commonly used units of capacity.

Force

Fasteners, like screws and rivets, exert force between the walls of two objects to hold them together. Force is measured in newtons (N). Force is exerted when making bends, folds, and cuts in sheet metal.

Pressure

Pressure is force applied over a unit area. Pressure is measured in pascals (Pa), or, more commonly, kilopascals (kPa). A pascal is defined as a force of one newton on a surface area of one square metre. Pressure may be exerted on gases, liquids, and even solids, and has many common applications.

PART ONE
Materials, Tools and Operations

1. SHOP SAFETY RULES

1. Report immediately to the instructor in the event of any accident, no matter how small.
2. Do not operate any machine until a complete demonstration has been given by the instructor.
3. If in doubt, ask for additional instructions. After you have been given the instructions, follow them.
4. Keep your hands away from your mouth and eyes when soldering.
5. Remember that the right way is the **best** way and the **safe** way.
6. When grinding, be certain that the glass shield is in place, or wear goggles.
7. Place tools and materials in their proper places so that they will not fall or cause injury.
8. Be certain that everyone is clear of the cutting blades of the squaring shear before stepping on the treadle.
9. Stop all machines before oiling, cleaning, or repairing them.
10. Turn off the power when you are finished with the drill press, the grinder, or other power machines.
11. Keep safety guards in place when using machines.
12. Keep your bench and floor clean. Place scrap metal in the proper receptacles, and leave nothing on the floor that could trip someone.
13. Be certain that chisels, rivet sets, prick punches, and other tools that may have burred heads are in proper condition before using them.
14. Do not use files without handles or tools with broken handles.
15. Report all unsafe conditions to the instructor at once.

Assignment

1. List eight safety rules.
2. Why are safety rules necessary?

2. COMMON SHEET METALS

Sheet Gauges. Metal thicknesses are designated by a series of numbers called **gauges**. Each gauge currently represents a decimal part of an inch. Metric gauge sizes have not yet been established. Sheets range in thickness from .006 in. to .125 in.; sheets over .125 in. are referred to as **plates**. As the sheet becomes thinner, the gauge number becomes larger.

The term "gauge" also refers to a tool used to check metal thicknesses. The two gauges in common use are:
1. U.S. Standard gauge, used for all ferrous metals, that is, metals containing iron.
2. Brown & Sharpe gauge, used for non-ferrous metals, such as aluminum, and for stainless steel.

Sheet Sizes. Thicknesses of metric sheets will probably range from 0.050 mm to 3 mm. Widths will probably run from 10 mm to 4000 mm, and lengths from 500 mm to 20 m. The most popular sizes for the sheet metal trade and school use will probably be 600 mm to 1000 mm in width, 2500 mm in length, and 0.30 mm to 1.00 mm thick.

Inch standard sizes are 30 in. and 36 in. wide and 96 in. and 120 in. long. Special sizes may be purchased at a slightly higher cost.

Base Metals

Base metals are metals which have been rolled into sheet form without the addition of other metals.

Hot Rolled Sheet Steel. These uncoated sheets have a low carbon content and look bluish-black near the sides and silver-grey around the centre. In many cases, articles are made of hot rolled sheet steel and later tinned, galvanized, painted, or enamelled. It is used for pans, cabinets, stove pipes, hoods, safety guards, tanks, etc.

Cold Rolled Steel. A further process gives a smoother, less porous, silver-grey finish to hot rolled steel sheets. This is desirable when a top-grade finish is needed for school lockers, cabinets, and furniture which will be painted.

Copper. This is a reddish-brown metal that is an excellent conductor of heat and electricity. It is one of the best roofing materials because its resistance to corrosion makes it possible to use lighter sheets. The metal is easily worked because of its malleability.

Aluminum. Because of its attractive appearance and its long-wearing and rust-resisting qualities, aluminum has become very popular in the sheet metal trade. Since it is approximately 2.5 times lighter than steel, it can be used where excessive mass would be a problem. Ducts, roofing, siding, trailers, and truck bodies are some of the many articles made from aluminum.

Coated Metals

These are base metals with a thin coating of another metal. Coating is usually done to protect the base metal from corrosion.

Electrolytic Tin Plate is a light gauge low carbon cold reduced steel coated with tin by electrolytic deposition. Tin plate thicknesses are indicated by symbols. For school use 1C and 1X are popular; 1X is thicker than 1C. For additional thicknesses 2X, 3X, and 4X are available.

The *base box* is a unit of area traditionally equivalent to 112 sheets, 14 in. by 20 in. in size. This base is used for pricing, as the amount of tin required for coating has been predetermined. Many schools use sheets 28 in. by 20 in. in size. Fifty-six of these sheets would equal a base box.

Many metric sizes will eventually be available. Sheets 700 mm by 500 mm will probably be used in schools.

Tin does not affect food and resists corrosion. These properties make it useful for articles such as kitchen utensils, food containers, and dairy equipment.

Galvanized Iron or Steel. This is the name given to sheet steel which has been coated with zinc to improve its appearance and prevent corrosion. Because of its zinc coating, galvanized iron is used for duct work in air conditioning and heating systems, and for cabinets, tanks, and refrigerators.

Terne Plate. An alloy of lead and tin makes up the coating on terne plate. Its main uses are for roofing and industrial fire doors as it is easy to solder and will withstand the effects of weather.

Alloys

Alloys are a combination of two or more metals. Rolling into sheet form brings out the best qualities of the metals.

Aluminum. Magnesium, manganese, and other metals are sometimes added to change the properties of aluminum for specific jobs, such as aircraft and mouldings.

Brass. This alloy is composed of copper and zinc in various properties. It tarnishes quickly but will not corrode. Brass is used extensively on water craft and wherever water might come in contact with the metal.

Monel Metal. The appearance of this alloy is similar to that of stainless steel, but monel metal is much easier to fabricate. It is composed of copper and nickel so that it is fairly expensive. Monel metal is non-corrosive and is used for store fronts and hospital and restaurant equipment.

Stainless Steel. The addition of chromium and nickel make this metal highly resistant to corrosion. Stainless is harder and higher in tensile strength than mild steel. It is used for many building components, such as kitchen, cafeteria, and restaurant installations; roof and drainage systems; and street front and entrance systems.

Assignment

1. What are sheet gauges?
2. When are sheets referred to as "plates"?
3. What does "non-ferrous" mean?
4. What is a base metal?
5. Name three base metals and give two uses for each.
6. Some metals are referred to as "alloys". What is an alloy?
7. Why are alloys necessary?
8. Why are some metals coated with another metal?
9. How are tin plate thicknesses indicated?
10. Give two uses for tin plate.
11. What is galvanized iron?
12. Name three articles around the home which are made from galvanized iron.
13. Describe the appearance of galvanized iron.

Properties of Metals

The properties of metals and their alloys are well marked. The different degrees in which these qualities are possessed by the different metals and alloys render each better adapted for certain purposes than others. Some of the more important properties are:

Metallic Luster—the degree to which the metal reflects light rays.

Tenacity—the tightness with which the atoms of the metal resist being pulled apart.

Ductility—the ease with which the metal can be drawn into wire.

Malleability—the ease with which the metal can be drawn out.

Conductivity—the degree to which a metal will transfer heat rapidly through its mass.

Fusibility—the state at which a metal fuses or passes from a solid to a liquid form.

Corrosiveness—the susceptibility of a metal to deterioration when exposed to the elements.

Hardness—the property which allows a metal to resist penetration by other hard objects.

Of the common metals, steel has the greatest tensile strength and gold and silver are the most malleable and ductile. Other important properties of metal are:

Relative Density—The mass of a metal compared to some standard, usually water, which is considered to have a relative density of 1.000.

Melting point—the temperature at which the metal changes from a solid to a liquid.

Metal	Relative Density	Melting Point
Aluminum	2.56	660°C
Zinc	7.1	420°C
Tin	7.3	230°C
Cast Iron	7.8	1200°C
Steel	7.8	1370°C
Copper	8.89	1080°C
Silver	10.4	960°C
Lead	11.4	330°C
Gold	19.3	1060°C

From the table above we can make a comparison with 1 l of water which has a mass of 1 kg. The same volume of lead would have a mass of $1 \times 11.4 = 11.4$ kg.

Assignment

1. List and define the following properties: tenacity, malleability, conductivity, ductility.
2. List the relative density and melting point of each of the following metals: aluminum, copper, steel, tin, lead.

3. HAND TOOLS

A journeyman is often judged by the quality and condition of his tools. The beginner should realize that, with proper use and care, tools will last longer, give better service, and be less expensive over a long period of time.

These are the hand tools most commonly used by sheet metal workers:

1. The *hack saw* is an adjustable saw which will cut metal. The blades may be replaced.
2. The *soldering iron* consists of a pointed piece of copper at the end of a steel rod which is held in a wooden handle. It is used to transfer heat from the furnace to melt solder during soldering operations.
3. *Combination snips* are so named because they can be used to cut metal along straight or irregular lines. They are used to cut the layout from the sheet metal.
4. The *try square* consists of a steel blade set in a steel beam at an angle of 90°. It is used to square short pieces of metal.
5. The *adjustable square* is used to obtain 45° and 90° mitres.
6. *Dividers* may be adjusted to scribe a circle of a given radius or to mark off given distances accurately on sheet metal.
7. *Aviation snips* have a compound lever action

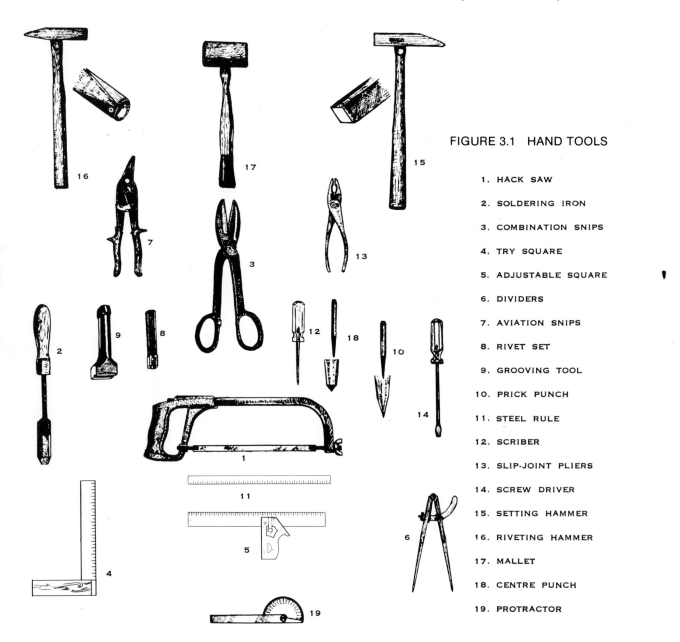

FIGURE 3.1 HAND TOOLS

1. HACK SAW
2. SOLDERING IRON
3. COMBINATION SNIPS
4. TRY SQUARE
5. ADJUSTABLE SQUARE
6. DIVIDERS
7. AVIATION SNIPS
8. RIVET SET
9. GROOVING TOOL
10. PRICK PUNCH
11. STEEL RULE
12. SCRIBER
13. SLIP-JOINT PLIERS
14. SCREW DRIVER
15. SETTING HAMMER
16. RIVETING HAMMER
17. MALLET
18. CENTRE PUNCH
19. PROTRACTOR

that facilitates the cutting of intricate patterns.
8. The *rivet set* is used to draw the sheets of metal and the rivets together and to form the head on the rivet. Rivet sets are available in a range of sizes to match the rivet sizes.
9. The *grooving tool* (or hand groover) is a steel tool used to lock a grooved seam by hand.
10. The *prick punch* is a steel rod which has been ground to a 30° angle at the tip. It is used to make indentations in sheet metal.
11. The *steel rule* is used for taking and laying out accurate measurements.
12. The *scriber* is a slender steel rod with a sharp point on one or both ends. It is used to scribe or scratch lines on a metal surface.
13. *Side-cutting pliers* can be used to grip small nuts to prevent them from turning and are also used for gripping and holding metal while soldering.
14. The *screwdriver* is available in many diameters and lengths and is made for either slotted head, Phillips head, or Robertson screws. Screwdrivers are used to drive in or remove screws.
15. The *setting hammer* has a square face and is used for setting down flanges and for wiring edges.
16. The *riveting hammer* has a rounded face and is used for riveting sheet metal.
17. The *mallet,* made of wood or rawhide, is used to form metal on stakes, since a metal hammer would damage the sheet metal surface.
18. The *centre punch* is similar to the prick punch, but its point is usually ground to an angle of 90°. It is used to mark the location of bend lines on heavy metal and the centres of holes to be drilled.
19. The *protractor* with a swinging blade is marked off in degrees and is used to measure and mark angles on sheet metal layouts.

Assignment

1. Make neat sketches of the following hand tools: riveting hammer, setting hammer, mallet, soldering iron, try square, combination snips.

4. BENCH STAKES

The worker can form sheet metal by bending it over stakes of various shapes. They vary in shape depending on the work to be done on them; square, round, and conical work can be formed, and edges and seams can be finished on them. Most stakes are made of a soft or cast steel.

Types of Bench Stakes

1. The *hollow mandrel stake* has a slot along its length in which a bolt slides. This permits the stake to be clamped firmly to the bench. Either the round or flat end can be used for forming, seaming, or riveting.
2. The *hatchet stake* is a sharp, bevelled stake with a hardened edge. It is used for making straight, sharp bends, for folding, and for bending edges.
3. The *square stake* has a flat, square head with a long shank and is used for general operations.
4. The *beakhorn stake* is a general-purpose stake. It has a round tapered horn on one end and a square tapered horn on the other. This stake is used for riveting, shaping round or square work, and for other general work.
5. The *candle mould stake* has two horns. The large-diameter horn is used for general purposes, while the small-diameter horn is used for tube forming or reshaping.
6. The *needle case stake* has a round, slender horn for small tubes and wire rings, and a heavier horn of rectangular cross-section for square work.
7. The *solid mandrel stake* has a double shank so that either the round or square edge can be used. This stake is used for forming or riveting square or rectangular work.
8. The *blowhorn stake* has two horns of different tapers. The apron end is used for shaping abrupt tapers; the slender tapered end is used for slightly tapered work.
9. The *bevel-edge square stake* has a square-shaped head with a bevelled edge. The shank is offset, thus allowing the stake to be used for a greater variety of jobs.
10. The *conductor stake* has two cylindrical horns of different diameters. It is used when forming, seaming, and riveting round objects.

FIGURE 4.1   BENCH STAKES

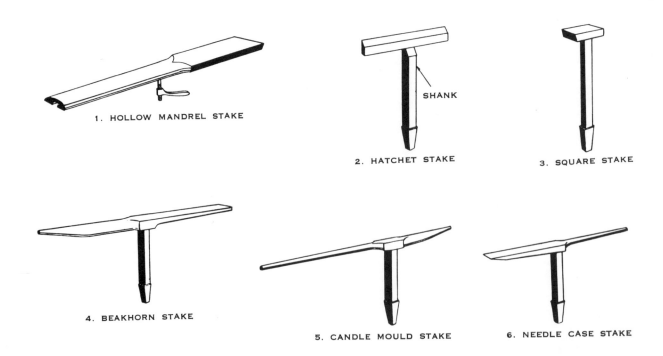

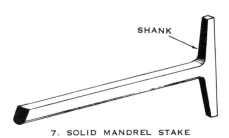
7. SOLID MANDREL STAKE

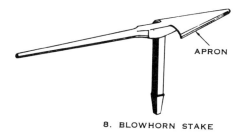
8. BLOWHORN STAKE

9. BEVEL-EDGE SQUARE STAKE

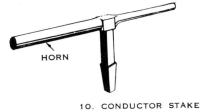
10. CONDUCTOR STAKE

Note: Safety Precaution:

Make sure stakes are placed in the proper openings in Bench Plate.

Assignment

1. From what material are stakes made?
2. What safety precaution must be observed when using bench stakes?
3. What is the purpose of bench stakes?
4. Draw neat sketches and label the following bench stakes: beakhorn, blowhorn, bevel edge, hollow mandrel.
5. Which stake would be best for general riveting on square or rectangular work?
6. Which stake is particularly useful for making straight sharp bends?

5. PROFILE VIEWS

A profile view is very similar to a sectional view, but does not show metal thickness. If you cut through a hollow metal object with a saw, the line on which the cut was made would be a profile view. Profile views are used frequently in the metal fabrication industry to develop cutting sizes and to show the shape and the direction of folds for the article being made. Profile views are usually drawn freehand and not to scale.

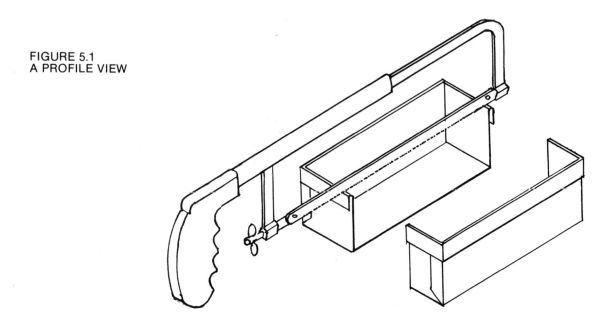

FIGURE 5.1
A PROFILE VIEW

FIGURE 5.2   PROFILE VIEWS

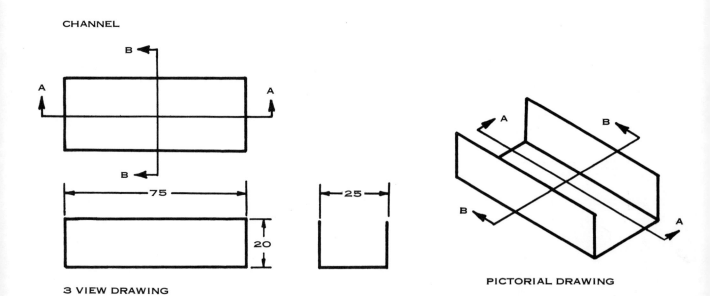

CHANNEL

3 VIEW DRAWING

PICTORIAL DRAWING

PROFILE VIEWS:

(1)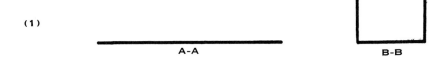

PROFILE VIEWS ARE USUALLY PLACED TOGETHER AND SHOWN AT RIGHT ANGLES TO EACH OTHER.

(2)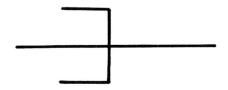

DIMENSIONS ARE SOMETIMES ADDED TO CALCULATE THE CUTTING SIZE.

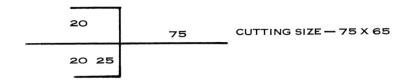

LAYOUT

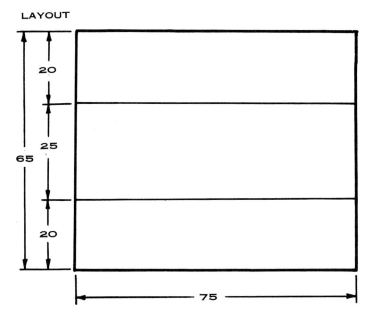

CUT MATERIAL TO SIZE AND DRAW LINES AS SHOWN IN PROFILE VIEW.

10 Sheet Metal Practice

RECTANGULAR BOX

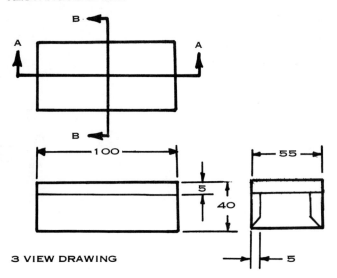

3 VIEW DRAWING

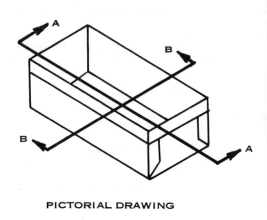

PICTORIAL DRAWING

PROFILE VIEWS

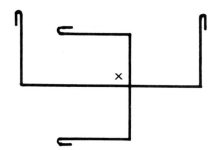

NOTE: 1. TABS ARE NOT SHOWN AS THEY DO NOT AFFECT THE CUTTING SIZE.

2. X—SIGNIFIES INSIDE OF BOX.

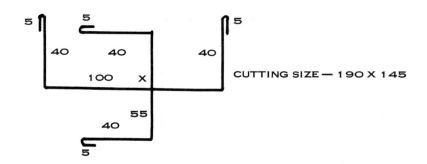

CUTTING SIZE — 190 X 145

LAYOUT

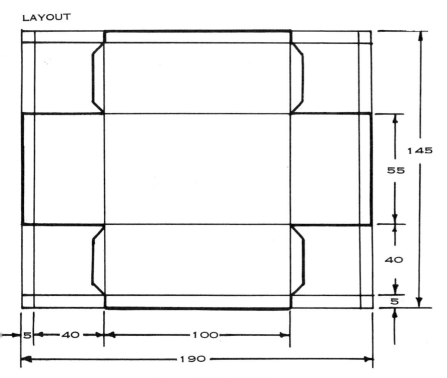

TABS ARE DRAWN TO COMPLETE LAYOUT.

Assignment

1. Explain the purpose of a profile view.
2. To what type of view is a profile view similar?
3. Why are tabs not shown in profile views?
4. What does the "x" signify on a profile view?
5. How are profile views usually placed and shown?

6. HAND OPERATIONS

SCRIBING A LINE

Accuracy is essential when laying out patterns.
1. Place a straight edge on the work in position for measuring.
2. Hold the scratch awl or scriber in the same way you would hold a pencil.
3. Press firmly on the rule with the fingers of the other hand.
4. Set the point of the scriber as close as possible to the end of the rule by tipping the top of the scriber outward.
5. Scribe a scratch mark by exerting pressure on the scriber. This mark should be made at both ends of the measurement on the material.
6. Place the scriber at one scratch mark and place one end of the straight edge against it.
7. With the scriber as a pivot move the straight edge until it lines up with the other scratch mark.
8. Scribe a line along the straight edge.

STRAIGHT CUTTING AND NOTCHING

Straight Cutting

1. Grasp the snips in one hand and the nearest part of the material in the other.
   NOTE: Rest the snips and material on the bench.
2. Open the blades as far as possible for your hand, and, using the throat of the snips, start the cut at the edge of the material.
3. To prevent leaving jagged edges, cut the material by closing the blades just short of their full length.
4. Start the snips at the extreme end of the preceding cut.
5. Finish the cut, keeping the snips on the line by changing the direction of the snips if necessary.

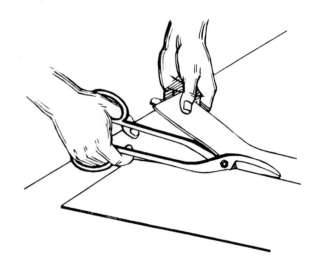

FIGURE 6.1 MAKING A STRAIGHT CUT

Notches and Their Uses

Notches are the spaces left when pieces of metal are cut from the pattern at the seams and edges. This is done to prevent surplus material from overlapping and causing a bulge at the seams and edges. Proper notching provides a better fit and gives the job a neater appearance. In some instances, it is merely necessary to slit an edge to allow the job to be formed.

While some specific uses of each notch are given, they can be used for other conditions. Sometimes a combination of notches is used.

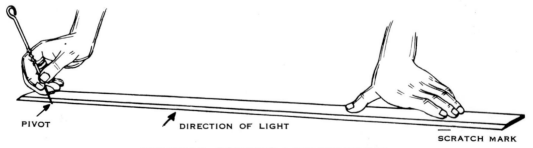

FIGURE 6.2 SCRIBING A STRAIGHT LINE

Hand Operations 13

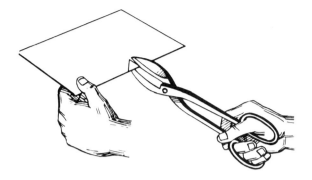

FIGURE 6.3 NOTCHING

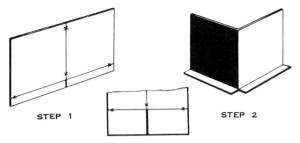

FIGURE 6.5 STRAIGHT NOTCH

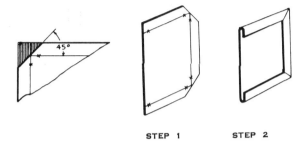

FIGURE 6.6 SLANT NOTCH

Notching. (Fig. 6.3)
1. Grasp the snips in one hand and the material in the other hand.
2. Open the blades; place the material between the blades.
   NOTE: *The points of the blades must not extend beyond the end of the notched line.*
3. Make the cut by closing the blades completely.
4. Cut the opposite side of the notch.

Square Notch. The square notch is square or rectangular in shape. The shaded portion shows the part to be cut away or notched (Fig. 6.4).

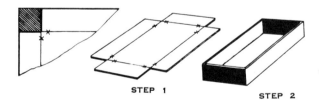

FIGURE 6.4 SQUARE NOTCH

This notch is used in the corners of boxes and pans to eliminate surplus material. Step 1 shows the notched pattern, and Step 2 shows the formed box with butt corners.

Straight Notch. A straight notch is a straight cut in the edge of the pattern or the job. (Fig. 6.5). It is used when dovetailing or from a flange at an outside corner. Step 2 shows the formed job.

Slant Notch. A slant notch is a cut made at an angle to the corner of the pattern (Fig. 6.6). It is cut at a 45° angle to prevent overlapping of the edges when single folds are to meet at right angles. This notch can be cut at other angles for various types of seams. Fig. 6.6, Step 1, shows a pattern notched for a single fold, and Step 2 shows the job with the edges folded flat and the sides of the notch meeting.

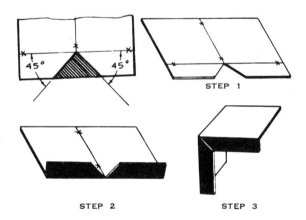

FIGURE 6.7 FULL "V" NOTCH

"V" Notch. The full "V" notch (Fig. 6.7) is a notch with both sides cut at a 45° angle to the edge of the pattern. The sides of the notch meet at a 90° angle. It is used when double seaming the ends of boxes and when making a job with a 90° bend and an inside flange.

Assignment

1. Explain how to scribe a line through two points.
2. When making a straight cut with combination snips, why are the blades not completely closed?
3. What part of the snips is used at the end of a cut?
4. At what angle should snips be held to the work while cutting?
5. Sketch neatly and name four types of notches.
6. Why are notches necessary?
7. How does a slant notch differ from a "V" notch?
8. What part of the snips is used to (a) notch (b) cut along a straight line?
9. Why is it necessary to notch accurately?

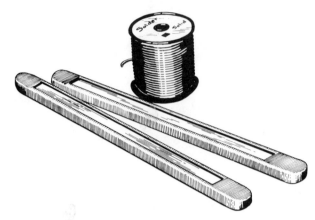

FIGURE 6.8 BAR AND WIRE SOLDER

SOLDERING

Solder

Soldering is a fast, simple method of joining two or more pieces of metal. The alloy adheres to metal sheets and melts at a lower temperature than the materials being joined.

Solder is one of the oldest, simplest, and most useful of all metal alloys; it was used by the Romans in the early years of the Christian era. It is generally classed as either "hard solder" or "soft solder".

Hard solder is harder and more difficult to apply than soft solder. It is usually applied with the use of a blowtorch or a welding torch. Two such solders are silver solder and bronze.

Soft solder is the solder most commonly used by sheet metal workers. It is a fusible alloy consisting mainly of tin and lead and is used to join two or more metals at temperatures below their melting points. Because of its low melting point, soft solder will flow readily and is easily applied with a soldering iron.

The most common solder alloy is approximately 50% tin and 50% lead by mass. These two metals combine to produce an alloy with a lower melting point than the base metals. For example, tin melts at 233°C and lead melts at 327°C but a solder alloy of 50% of each will melt at 213°C.

Solder is available in either bar or wire form. Bar solder is made in various shapes. Most wire solder is wound on spools. This solder ranges in diameter from 0.3 mm to 6.5 mm and usually consists of 60% tin and 40% lead. Wire solder is also obtainable with either an acid or a rosin core. The acid or rosin is a flux, and no other flux need be applied from other sources.

NOTE: When referring to percentages of tin and lead, tin is always mentioned first.

No tin-lead solder except the "eutectic" (62% tin—38% lead) has what is commonly called a melting "point". All others have a melting range, extending from the temperature shown on the lower boundary line of the chart to that shown at the higher line. At any temperature within this range the solder is a mixture of liquid and solid and is "pasty" or "plastic".

An example of this characteristic in solders would be the widely used 30/70 alloy which begins to melt at 183°C, has a plastic range of 75°C, and is finally all liquid at 258°C.

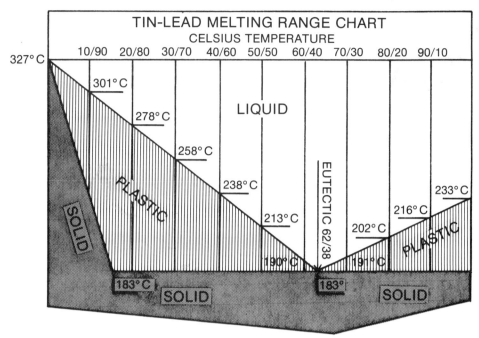

FIGURE 6.9

Assignment

1. Explain the process of soldering.
2. What are the two general classifications of solder?
3. What type of solder is used most often by the sheet metal worker?
4. Of what two metals is soft solder composed?
5. On what does the melting temperature of soft solder depend?
6. Does half-and-half solder have equal proportions of two metals by mass or by quantity?
7. At what temperatures does half-and-half (50/50) solder melt?
8. What happens to the melting temperature of soft solder as more lead is added?
9. Explain what the following expressions mean when related to soft solder: (a) 50/50 (b) 60/40 (c) 40/60 (d) 30/70.
10. At what temperature does (a) tin (b) lead melt?

Soldering Fluxes

A flux is a substance used to clean the metal to be soldered and to prevent an oxide from forming on the material while soldering.

There are two types of fluxes: corrosive and non-corrosive. Muriatic acid is an example of a corrosive flux, and rosin is a non-corrosive flux. It is necessary to remove all corrosive fluxes after soldering to prevent any further action of the flux on the metal.

NOTE: For a better appearance and as a safety precaution, it is advisable to remove all fluxes after soldering, whether they are corrosive or non-corrosive.

Metal	Flux Required
Tin plate	Soldering paste or rosin
Galvanized iron	Muriatic (hydrochloric) acid
Copper, brass	Chloride of zinc, soldering paste, or rosin
Steel	Pure chloride of zinc or a mixture of equal parts of sal-ammoniac and borax.
Stainless steel	Phosphoric acid or chloride of zinc.

NOTE: Metal should be etched approximately one minute with muriatic acid before applying chloride of zinc.

Soldering paste is manufactured by suspending a chloride solution in a grease or petrolatum. As a flux, it is used principally for clean tin plate.

Muriatic acid is a yellow solution of hydrogen chloride, which comes from sea salt, and water.

Sheet Metal Practice

This solution is an impure form of hydrochloric acid.

Chloride of zinc (killed or cut acid) is made by dissolving as much zinc as possible in muriatic acid. The solution should be kept in an earthenware or glass jar as it has a corrosive action.

Rosin is the amber-coloured chemical compound that remains after the turpentine has been removed from the sap of certain pine trees. It does not clean the surface of the work, but it prevents oxidation during soldering by covering the surface with a film.

Assignment

1. What is the purpose of a flux?
2. At what time during the soldering process are fluxes used?
3. Explain the term "corrosive".
4. Why is it necessary to use a corrosive flux?
5. Why should fluxes be washed off the metal immediately after the soldering operation?
6. Outline the composition of soldering paste.
7. On what metal is muriatic acid used?
8. Is muriatic acid a corrosive or a non-corrosive flux?
9. What is the technical name for muriatic acid?
10. Give another name for zinc chloride and name three metals on which it is used as a flux.

Soldering Irons

Soldering irons are sometimes referred to as *soldering coppers.* The size is specified as the mass per pair of coppers.

The soldering iron is used to convey heat from the fire to the place to be soldered. Copper is a good conductor of heat as it absorbs heat rapidly. However, when heated, the copper unites with the oxygen in the air to form copper oxide. Since this oxide is a poor heat conductor, it must be eliminated by coating the point of the soldering iron with solder. This is known as "tinning" a soldering iron.

Tinning a soldering iron using sal-ammoniac

1. Clamp the soldering iron in a vice and file the point clean.
2. Heat the copper until it will melt solder easily.
3. Rub the side of the copper to be tinned on a block of sal-ammoniac.
4. Melt a few drops of solder on the sal-ammoniac and continue the rubbing motion until the point is coated with solder or "tinned".

NOTE: After the soldering iron has been tinned, it may be kept clean by dipping the point in a solution of one part of sal-ammoniac to forty parts of water. The solution should be kept in a glass or earthenware vessel as it has a corrosive action. On occasion a damp cloth may be used to wipe the point of the copper clean.

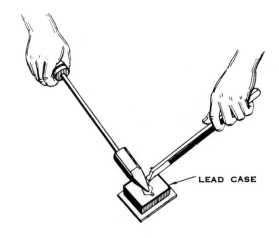

FIGURE 6.11 TINNING WITH SAL-AMMONIAC

FIGURE 6.10 SOLDERING IRON

Tinning a soldering iron using rosin

1. Heat the copper until it will melt solder freely.
2. Clamp the copper in a vice and file the point clean.
3. Reheat the soldering iron if necessary.
4. Melt a few drops of solder on the rosin.
5. Rub the point of the soldering iron in powdered rosin until the point is tinned.
6. Clean the tinned surface of the copper.

FIGURE 6.12 TINNING WITH ROSIN

Assignment

1. Name three parts of a soldering iron and note the material from which each part is made.
2. Give another name for a soldering iron.
3. What is the purpose of a soldering iron?
4. Why is a soldering iron tinned?
5. Explain how to tin a soldering iron using
   (a) rosin
   (b) sal-ammoniac.
6. What purpose does sal-ammoniac serve when tinning a soldering iron?
7. What is the purpose of a dipping solution?
8. Explain how to prepare a dipping solution.
9. What precautions must be taken when tinning a soldering iron?
10. Why is the head of the soldering iron made of copper?

Gas Furnaces

The best method of heating soldering irons in the shop is with a gas furnace consisting of a gas stove with a hood. The hood can be raised to facilitate cleaning of the furnace bed. A refractory lining inside the furnace prevents the metal part of the furnace from being burned. Both the hood and the lining are shaped to deflect the heat over the soldering irons.

Most furnaces are equipped with two burners controlled by separate valves that can be adjusted to the desired temperature. The furnace also has a pilot light which is allowed to remain burning during the working period when the furnace is not in use.

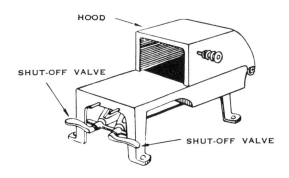

FIGURE 6.13 GAS FURNACE

Soldering a Lap Joint

1. Heat the soldering iron in the gas furnace. Tin the iron if necessary.
2. Place the joint to be soldered in the proper position.
3. Apply the proper flux on the joint with a swab.

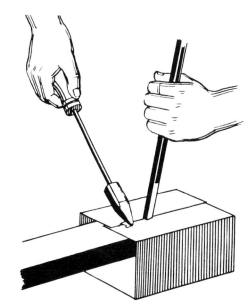

FIGURE 6.14 TACKING A JOINT

18 Sheet Metal Practice

4. Clean the point of the soldering iron. Tack each end of the joint.
5. Starting at the far end of the joint, place one face of the copper flat on the metal until the solder starts to flow freely into the joint.

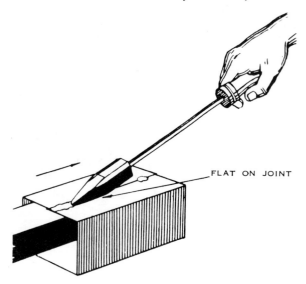

FIGURE 6.15 SOLDERING A JOINT

6. Draw the soldering iron very slowly along the joint, and add as much solder as is necessary without raising the soldering iron from the joint.
7. Remove the excess flux from the metal.

NOTE: Heat the soldering copper as often as necessary to retain a uniform sweating effect throughout the joint.

Assignment

1. What element unites with copper to form copper oxide?
2. When a lap joint is being soldered, why is it tacked first?
3. Explain two ways to clean the point of a soldering iron.
4. Why should the soldering iron be placed over the top piece of metal when soldering a lap joint?
5. Why should the soldering iron be drawn slowly over the joint?
6. How is the flux applied to the joint?
7. Draw a neat sketch showing a joint being (a) tacked (b) soldered.
8. Explain how to heat a soldering iron.
9. Explain, in point form, how to solder a lap joint.
10. How does soft soldering differ from welding?

HAND GROOVING

A grooved seam consists of two folded edges hooked together and offset. There are two ways of making this seam — the hand method and the machine method. The latter is used when the production of many pieces is required. The grooved seam is used for round pipe and containers.

FIGURE 6.16 STEPS IN MAKING A GROOVED SEAM

Grooved Seam Allowance

For material lighter than 0.5 mm, three times the width W of the grooved joint is added to the pattern for the lock, half of this being added to each end of the pattern. (Fig. 6.16). The folds are full and opposite.

Grooved seams are made by hooking together, offsetting, and flattening two folded edges. The seams are offset to the outside of the job as illustrated in Fig. 6.16. The *hand groover* is used to offset an outside grooved seam (Fig. 6.17). Grooving tools currently come only in fractional inch widths. Metric groovers were not available at the time of publication. For inch groovers, each width is designated by a number. The following table shows the width of seam and allowance necessary for each number. In order to aid students working in millimetre dimensions, approximate metric equivalents are given in brackets. *Remember that these dimensions are only soft conversions and do not represent the true seam size of inch designed hand groovers.* Metric groovers, should they eventually be made available, may come in different sizes than the dimensions in brackets. They may also be designated by an entirely different means than the number system used for inch groovers.

Number of Groover	Width of Seam	Total Allowance	Amount Each End	Amount Folded Each End
	W	3W	1½ W	W
5	⅛ in. (3 mm)	⅜ in. (9 mm)	3⁄16 in. (4.5 mm)	⅛ in. (3 mm)
4	3⁄16 in. (5 mm)	9⁄16 in. (15 mm)	9⁄32 in. (7.5 mm)	3⁄16 in. (5 mm)
3	¼ in. (6 mm)	¾ in. (18 mm)	⅜ in. (9 mm)	¼ in. (6 mm)
2	5⁄16 in. (8 mm)	15⁄16 in. (24 mm)	15⁄32 in. (12 mm)	5⁄16 in. (8 mm)
1	⅜ in. (9 mm)	1⅛ in. (27 mm)	9⁄16 in. (13.5 mm)	⅜ in. (9 mm)

Finishing the Grooved Seam

1. Fold the edges to the proper width and form the job to shape.
2. Hook the folds together.
3. Place the article on a suitable stake.
4. Flatten the seam slightly with a mallet.
5. Place the hand groover over one end of the seam and strike it with a hammer.
6. Groove the other end in the same manner (Fig. 6.18).
7. Groove the entire seam by striking the hand groover with the hammer, while moving the groover along the seam.

FIGURE 6.17 THE HAND GROOVER

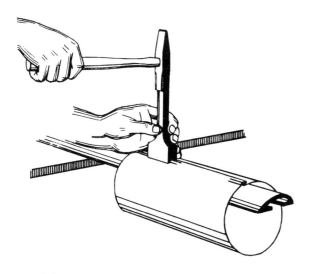

FIGURE 6.18 GROOVING THE SEAM

Assignment

1. Explain the purpose of a grooved seam.
2. On what does the width of a grooved seam depend?
3. What type of fold is made on each end of the pattern for a grooved seam?
4. When are the folds for the grooved seam made?
5. Draw a neat sketch of the end section of a grooved seam.
6. What is the total allowance for a grooved seam?
7. How much of the total allowance is placed on each end of the pattern?
8. How much of the allowance is folded on each end of the pattern?
9. What does the amount that is folded on each end of the pattern equal?
10. In which direction are the folds made for a grooved seam?

WIRING

The Wired Edge

The edges of some jobs are formed around wire to strengthen and stiffen them and to eliminate sharp edges. Wired edges can be made by either hand or machine. The hand process is used when a machine is not available or when the work is too large, the material too heavy, or the shape not suitable for the capacity of the machine available.

FIGURE 6.19   THE WIRED EDGE

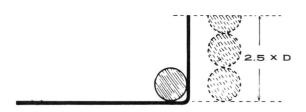

FIGURE 6.20   MATERIAL REQUIRED FOR A WIRED EDGE

Wire may be made of iron, steel, copper, aluminum, stainless steel, nickel, and other metals. Iron wire is very soft and is seldom used; annealed steel wire is stronger and is the type used by the sheet metal trade. Hard drawn (not annealed) steel wire, which is very strong, can be obtained for special purposes. Iron and steel wire may be obtained plain or coated with tin, copper, zinc, or other metals. Coating reduces the tendency of wire to corrode and makes the wire easier to solder.

Wire diameters are currently expressed in fractions of an inch, millimetres, or by a series of gauge numbers corresponding to the diameter of the wire in decimal parts of an inch. While the American steel and wire gauge is presently used for iron and steel wire, there are so many series of gauge numbers for the various kinds of wire that, in order to avoid mistakes, it is best to order wire by fractional inch, decimal inch, or millimetre sizes. Wire which exceeds a certain diameter is called rod. By current standards, wire larger than 6 mm or .25 in. is designated as rod.

FIGURE 6.21   FORMING THE METAL OVER THE WIRE

Wiring a Straight Edge

1. The amount of material required for wiring a straight edge is 2.5 times the diameter of the wire. (T.A. = 2.5 × D) Fig. 6.20.
2. Bend or fold the allowance to 90°. Place the wire in the folded edge and form the metal over the wire using a mallet (Fig. 6.21). *NOTE: When wiring, it is easier to work from right to left. The wire is held tightly in place with the left hand; however, if this procedure proves difficult, pliers may be used.*

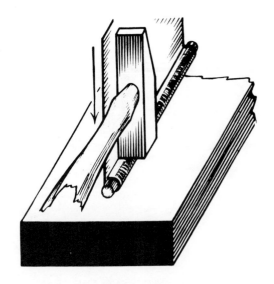

FIGURE 6.22   CLOSING THE METAL TO THE WIRE

3. Close the metal further to the wire by striking the allowance with the setting hammer (Fig. 6.22). *NOTE: A piece of scrap metal could be placed against the side of the article to prevent damage to the face of the metal.*
4. Use the peening end of the setting hammer to complete the wiring operation. (Fig. 6.23)

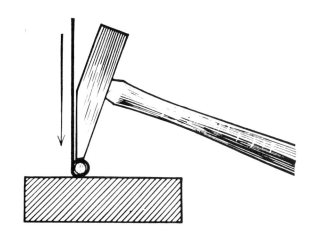

FIGURE 6.23 COMPLETING THE WIRED EDGE

Assignment

1. State the allowance for a wired edge.
2. Make a neat sketch to show the allowance for a wired edge.
3. What is the purpose of a wired edge?
4. Name four places where wire edges are used.
5. Why is a mallet used to start forming the metal over the wire?
6. Which hammer is used to complete a wired edge?
7. Explain how to complete a wired edge.
8. If the wired edge allowance is known, how is the diameter of the wire determined?
9. If the diameter of wire is known, how is the wired edge allowance determined?

RIVETING

Riveting is one of the most important methods used to join pieces of metal permanently. Riveted joints are used where strength is needed and when metals of heavy gauges must be joined. Hand riveting is done with a hammer and a rivet set.

Types of Hand Rivets

The *tinners' rivet* and the *flat-head rivet* are the most common types. The body of the rivet is called the "shank". The length and diameter of each rivet is measured as shown in Fig. 6.24. The head of the tinners' rivet is thinner and of a larger diameter than that of the flat-head rivet.

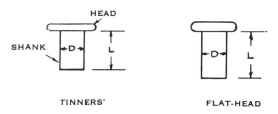

FIGURE 6.24 TYPES OF RIVETS

Tinners' rivets are currently designated by the mass, in *inch-pound* units, of 1000 rivets. No standards for designating rivets in metric terms were available at the time of publication. By the *inch-pound* standards, 8 oz. rivets have a mass of 8 oz./1000; 1 lb. rivets have a mass of 1 lb./1000; and 2 lb. rivets have a mass of 2 lb./1000. As the rivets increase in mass, the diameter and length become greater. Flat-head rivets vary in diameter from 3/32 in. to 7/16 in. in 1/32 in. steps. For students working to millimetre dimensions approximate equivalents of these diameters are from 2.4 mm to 11.1 mm in steps of 0.8 mm. (Again, remember that metric rivets, when they become available will likely come in different sizes than the soft-converted figures given here). Rivets of these diameters can be obtained in different lengths; longer rivets are used when more than two thicknesses of sheet metal or pieces of heavier metal are to be riveted together.

Rivet Sets

A rivet set is made of tool steel. The large end has a deep hole and a shallow, cup-shaped hollow (Fig. 6.25). The deep hole fits over the rivet and is

22 Sheet Metal Practice

used to draw the sheets and the rivet together. The cup-shaped hollow is used to form the head on the rivet. A rivet set can be used to draw rivets directly through thin metal without previously punching a hole. The drawing hole has an outlet at the end to allow the burrs to drop out.

Rivet sets are available for each different size of rivet. A rivet set should be selected with a hole slightly larger than the diameter of the rivet.

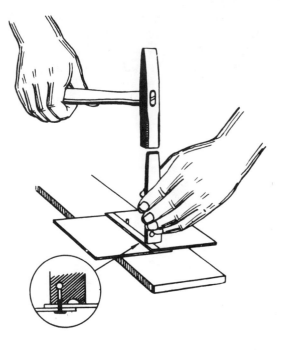

FIGURE 6.26   DRAWING THE RIVET AND METAL SHEETS TOGETHER

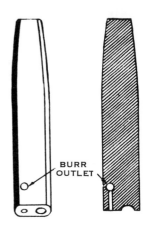

FIGURE 6.25   RIVET SET

Steps in Riveting

1. Place the job to be riveted on a suitable plate or bench stake.
2. Select the rivet, the riveting hammer, and the proper rivet set.
3. Insert the rivet in the hole, resting the head of the rivet on the stake.
4. Draw the material and the rivet head together tightly by placing the hole in the rivet set over the rivet and striking the set one or two sharp blows with the hammer (Fig. 6.26).
5. Place the cup-shaped hole on the rivet, and strike the rivet set one or two sharp blows with the hammer (Fig. 6.27). This is done to head the rivet.

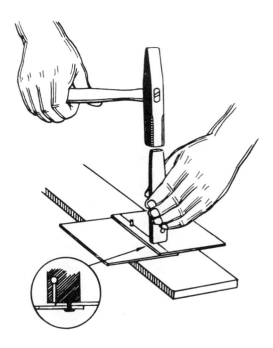

FIGURE 6.27   HEADING THE RIVET

Blind Riveting

Blind riveting is so named because the rivet can be set from one side of the work. Unlike many other fasteners which require access to both sides of the work, the blind rivet has emerged over the past few years from the category of a special fastener to become the preferred fastener for an endless variety of applications, even when access to both sides of the work exists.

Generally blind rivets are classified by the method in which they are set. Pull-stem, drive-pin, and chemically expanded are the three basic types.

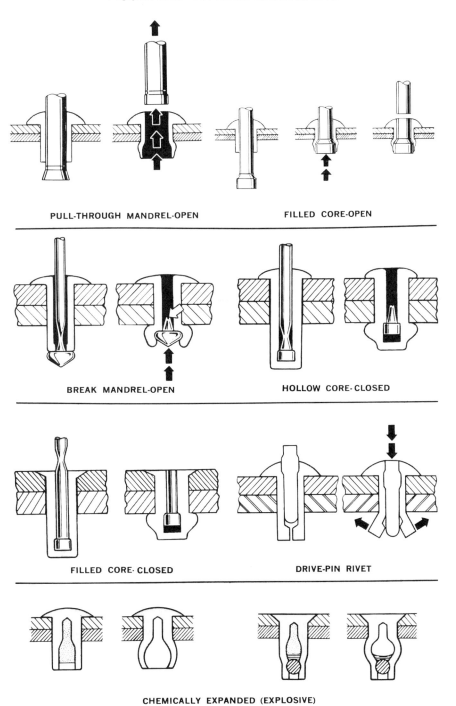

FIGURE 6.28 TYPES OF BLIND RIVETS

PULL-THROUGH MANDREL-OPEN FILLED CORE-OPEN

BREAK MANDREL-OPEN HOLLOW CORE-CLOSED

FILLED CORE-CLOSED DRIVE-PIN RIVET

CHEMICALLY EXPANDED (EXPLOSIVE)

24 Sheet Metal Practice

Pull-stem rivets can be subdivided into pull-through, self-plugging, and break-mandrel rivets. In the latter type a part of the mandrel remains as a plug in the hollow rivet body. This type of riveting is done with tools which are operated manually, hydraulically, or pneumatically.

FIGURE 6.29   TYPES OF PULL-STEM RIVETERS

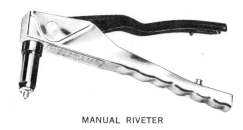

MANUAL RIVETER

PNEUMATIC RIVETER

AIR-HYDRAULIC RIVETER

The purpose of the blind riveting tool is to pull a high strength mandrel into the rivet. The mandrel head expands and sets the rivet, drawing the parts of the work tight. At the proper point of tension and not before, the mandrel breaks, leaving the rivet set in place.

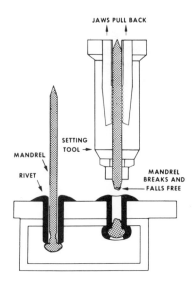

FIGURE 6.30   SETTING A PULL-STEM RIVET

Pull-stem rivets insure tight, strong durable joints regardless of application or operator skill. The mandrel breaking point combined with the force required to set the rivet runs as high as 12 500 N. This results in a 2670 N squeeze between parts.

Drive-pin rivets have pins which are hammered into the rivet body, causing the slotted ends of the rivet to flare out on the blind side of the work.

Chemically expanded explosive rivets have an explosive charge in the body which, when activated by heat or electrical charge, expands and sets the blind rivet end.

Blind rivets are made of various materials, including steel, copper, stainless steel, aluminum, and monel metal. Steel rivets are used where high strength with minimum corrosion resistance is required, while aluminum rivets are commonly used for exterior work or where resistance to corrosion is needed. Monel and stainless steel rivets can be used where both high strength and corrosion resistance are factors.

Currently, blind rivets are available only in nominal inch sizes. In mechanically expanded types, 3/32 in., 7/64 in., 1/8 in., 5/32 in., 3/16 in., 1/4 in., and 5/16 in. are most common. Explosive rivets are produced oversize, with the actual rivet diameter about 1/64 in. larger than the diameter specified. Nominal inch sizes are 1/8 in., 5/32 in., and 3/16 in.

Assignment

1. Name the parts of a rivet.
2. When are riveted joints used?
3. Name the two common types of rivets.
4. How are the sizes of tinner's rivets determined?
5. Why is there an outlet on the side of each rivet set?
6. What is the purpose of the deep hole in the bottom of each rivet set?
7. Why is the shallow cup-shaped hollow necessary in the bottom of the rivet set?
8. Make neat sketches showing the riveting operation.
9. List three types of blind rivets.
10. State an advantage blind riveting has over regular riveting.

PUNCHING

The Hand Lever Punch

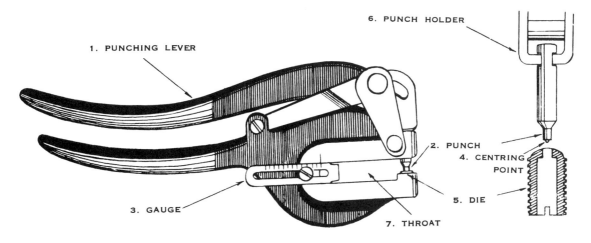

FIGURE 6.31   THE HAND LEVER PUNCH

The hand lever punch is used for punching small holes near the edges of light sheet metal. It consists of a die and a punch which is moved by a lever. The punch forces the metal through the die, leaving a clean hole. Each punch is furnished with corresponding punches and dies of various sizes.

Principal Parts

1. The *punching lever* activates the punch holder and, therefore, the punch.
2. The *punch*.
3. The *gauge* enables the operator to punch more than one hole at the same distance from the edge of the metal.
4. The *centring point* locates the centre of the hole to be punched.
5. The *die* is threaded and has a slot in the lower end to facilitate changing it.
6. The *punch holder* has flanges which fit into the recesses of the punch.
7. The *throat* governs the distance from the edge of the metal to the hole to be punched.

Changing the Punch and Die

1. Remove the die with a screw driver, and use the screw driver to remove the knurled, slotted screw that holds the punch holder in place.

FIGURE 6.32   REMOVING THE DIE

26 Sheet Metal Practice

2. Raise the punching lever. Slide the punch holder back far enough to release the punch.
3. Hold the hand lever punch and remove the punch.
4. Insert the required size of punch with the recesses parallel to the flanges of the punch holder.
5. Bring the levers together, being certain that the flanges on the punch holder slide into the grooves of the punch (Fig. 6.33).

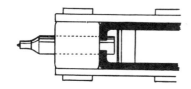

FIGURE 6.33   REPLACING THE PUNCH

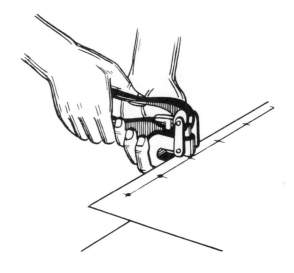

FIGURE 6.34   PUNCHING HOLES

6. Replace the knurled, slotted screw at the centre of the hand lever punch. Insert the proper size of die.
7. Adjust the die until the end of the punch enters the die about 1.0 mm to 1.5 mm when the levers are closed.

Punching Holes

1. Lay out the holes in the required places on the pattern and mark the centres with a prick punch or centre punch.
2. Check to see if the correct size of punch is in the hand lever punch. If it is not, change the punch and die.
3. Lay the sheet on the bench with the marked holes extending beyond the edge of the bench.
4. Place the centring point of the punch in the centre mark on the sheet (Fig. 6.34)
5. Hold the throat and the lower lever in a horizontal position.
6. Punch the hole by pressing the upper lever down.

Assignment

1. What is the capacity of the hand lever punch?
2. Explain how to change a punch and die.
3. Make a neat sketch of a hand lever punch and label its parts.
4. Why is an indentation necessary when using the hand lever punch?
5. Which hand tool is used for making an indentation in 0.40 mm galvanized iron?
6. Explain, in point form, how to punch a hole.
7. State the function of the gauge.
8. How far should the punch extend into the die?
9. What are the limitations of the hand lever punch?
10. Why is a hand lever punch sometimes preferable to a drill press?

7. MACHINES AND MACHINE OPERATIONS

THE FOOT OPERATED SQUARING SHEARS

Squaring shears are used to trim, cut, and square sheet metal. Trimming and cutting can be done on marked sheets or by using the gauges provided on the machine. The latter method is used when many pieces of the same size are required. Only one sheet should be cut at a time. Under no circumstances should wire, rod, bar stock, or seamed or grooved edges be cut on the squaring shears, since these could nick the cutting blade.

Principal Parts

1. The *crosshead* supports the top blade.
2. The *bottom blade* is attached to the bed.
3. The *bed* supports the metal to be cut. The *graduations* on the bed simplify setting of the front gauge.
4. The *foot treadle* operates the cutting blade.
5. The *front gauge* adjusts the machine to cut identical pieces in a series.
6. The *side gauge* is adjusted to square sheets on the machine.
7. The *bevel gauge* is set to make the machine cut sheets at a given angle.
8. The *safety guard*.
9. The *hold-down handle* operates the *hold-down* which keeps the sheets in place.

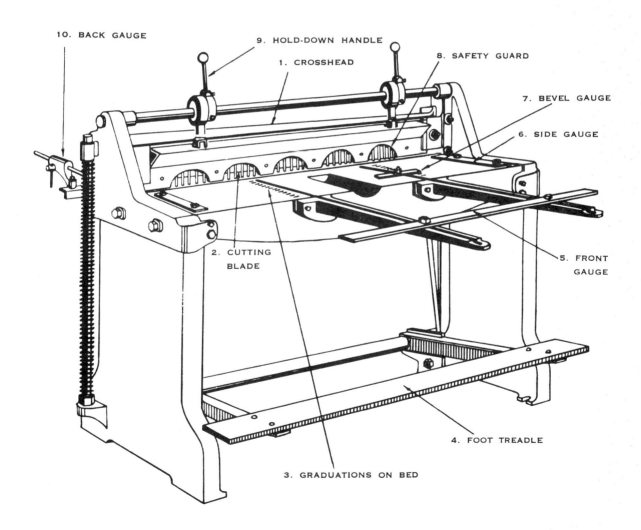

FIGURE 7.1 FOOT-OPERATED SQUARING SHEARS

10. The *back gauge* is adjusted to allow the operator to cut a series of narrow pieces.

NOTE: SAFETY PRECAUTION:

Keep your hands and feet clear of the machine before stepping on the foot treadle.

Securing a Working Edge

When the layout is square or rectangular, it is necessary to have a working edge.
1. Select the longest straight edge on the piece of material and place that edge against either side gauge on the bed of the squaring shears.

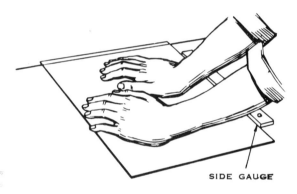

FIGURE 7.2 POSITIONING HANDS WHEN SECURING A WORKING EDGE

2. Slide the material between the blades so that the metal extends about 3 mm beyond the lower blade.
3. Hold the sheet firmly against the side gauge, then step on the foot treadle.
4. Make the cut at a 90° angle from the straight edge, and remove the material.

Assignment

1. What operations can be performed on the squaring shears?
2. State the capacity of the squaring shears.
3. What safety precaution must be observed while using the squaring shears?
4. What supports the top blade of the squaring shears?
5. How is the hold down engaged with the metal?
6. Why is a side gauge required?
7. What do the graduations on the bed of the squaring shears indicate?
8. How many sheets of metal should be cut at a time?
9. Explain two methods of cutting and trimming.
10. Why should wire, rod, and bar stock not be cut on the squaring shears?
11. How is a working edge prepared?
12. Why is a working edge necessary?

THE BAR FOLDER

The bar folder is used to fold the edge of a sheet to a desired angle, for making channel shapes (double right angle folds), and for preparing folds for locked seams and wired edges. While narrow channel shapes can be formed, reverse folds cannot be made too close together.

Principal Parts

1. The *operating handle* turns the wing.
2. The *wing* governs the sharpness or roundness of the fold.
3. The *45° and 90° stops* set the machine to make a 45° or 90° fold.
4. The *folding blade* is the part around which the metal is folded.
5. The *gauge-adjusting screw* adjusts the machine for the width of fold required.
6. The *adjustable collar* adjusts the machine to make folds at special angles.
7. The *gauge-locking screw* holds the gauge in place.
8. The *fingers* act as a stop for the sheet of metal.
9. The *wedge lock nut* clamps the wing in place.
10. The *wedge-adjusting screw* adjusts the wing.

NOTE: SAFETY PRECAUTION:

Keep your fingers away from the folding blade when operating this machine.

Machines and Machine Operations   29

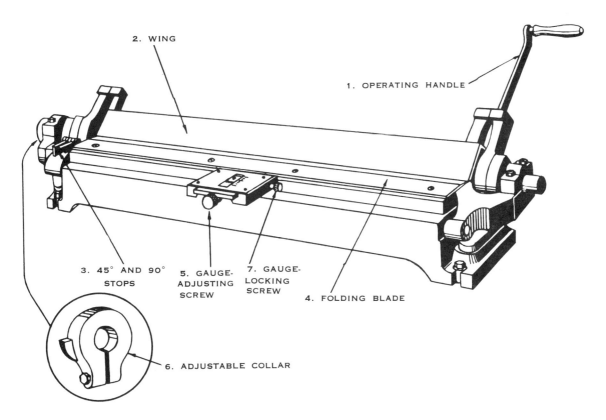

FIGURE 7.3   THE BAR FOLDER — FRONT VIEW

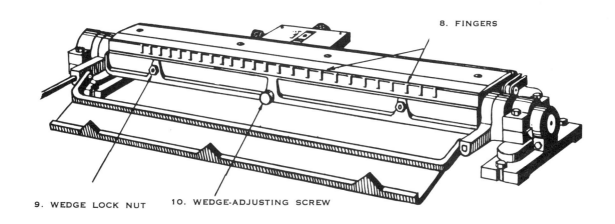

FIGURE 7.4   THE BAR FOLDER — REAR VIEW

Forming a Single Fold

1. Set the gauge, by means of the adjusting screw, to the desired fold width.
2. Insert the edge of the metal to be folded between the folding blade and the jaw.
3. Hold the metal firmly against the gauge fingers with the left hand and place the right hand on the operating handle. (Fig. 7.5).

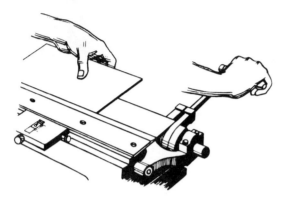

FIGURE 7.5   STARTING THE FOLD

4. Pull the operating handle with the right hand as far as it will go to fold the edge. (Fig. 7.6). *NOTE: Keep the left hand on the sheet until the sheet is clamped.*

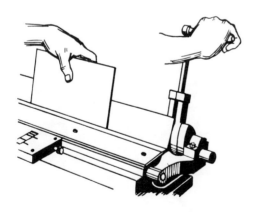

FIGURE 7.6   MAKING THE FOLD

5. Return the operating handle to its former position. *NOTE: Keep your hand on the operating handle until the wing is back to its normal position.*
6. Remove the sheet from the folder. Place it with the fold facing upwards on the bevelled part of the blade, as close as possible to the wing (Fig. 7.7).

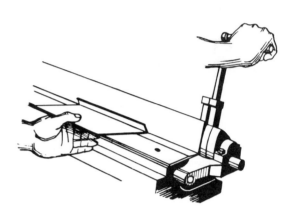

FIGURE 7.7   FLATTENING THE FOLD

7. Pull the operating handle with a swift motion to flatten the fold. This completes the fold.
8. Return the operating handle to its former position.

Types of Folds

Many types of folds can be made on the bar folder. The following illustration shows some common applications.

FIGURE 7.8 TYPICAL FOLDS FORMED ON THE BAR FOLDER

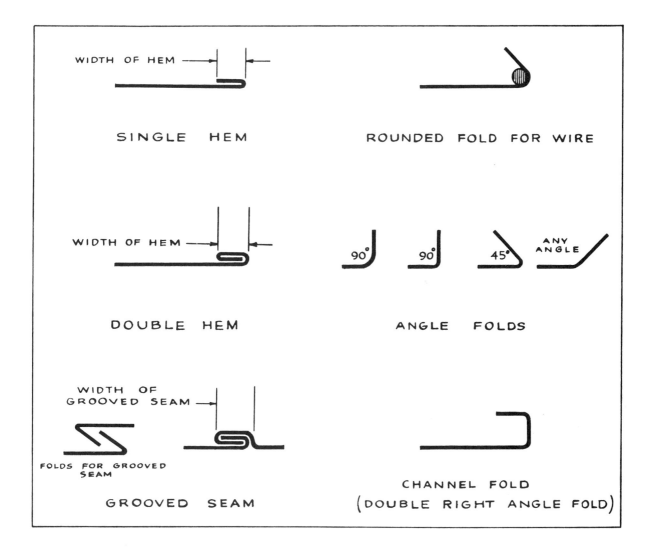

Assignment

1. When are the two stops on the bar folder used?
2. What operations can be performed on the bar folder?
3. Sketch three folds that can be made on the bar folder.
4. State the capacity of the bar folder.
5. What safety precaution must be observed while using the bar folder?
6. How much does each graduation represent on the gauge adjustment?
7. What is the purpose of the fingers on the bar folder?
8. What do the graduations show?
9. How would you set up for a fold of 75° on the bar folder?
10. What part must be loosened before adjusting the wing?

THE STANDARD HAND BRAKE

The standard hand brake makes sharp and rounded angle bends. Curved shapes can also be formed by using attachments called moulds. Material heavier than the capacity specified by the manufacturer should *not* be formed in the brake. Never bend wire, rod, or band iron, as this will damage the upper jaw blade and the bending leaf.

Principal Parts

1. The *bending leaf* is the part of the machine which forms the metal.
2. The *upper jaw* or *clamping bar* grips the metal and holds it in place.
3. The *lower jaw* supports the metal during the braking operation.
4. The *bending leaf handle* operates the bending leaf.

FIGURE 7.9 STANDARD HAND BRAKE

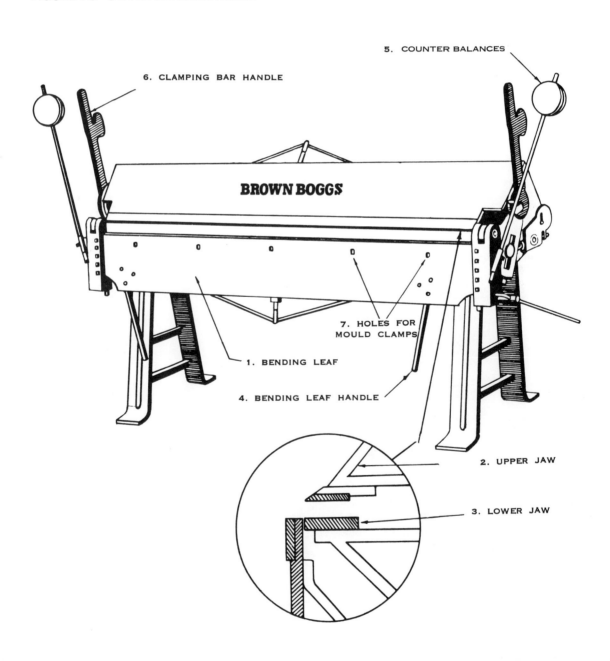

5. The counterbalances facilitate the braking operation by counterbalancing the weight of the bending leaf.
6. The *clamping bar handle* moves the upper jaw.
7. The *holes for mould clamps* are used to attach the mould clamps to the machine.

NOTE: SAFETY PRECAUTION:

Keep fingers away from the bending leaf and clamping bar.

Making a Right-Angle Bend

1. Place the material with the centre of the bend line on the left end flush with the edge of the clamping bar (Fig. 7.10).

FIGURE 7.11   CLAMPING THE RIGHT SIDE OF A SHEET

FIGURE 7.10   CLAMPING THE LEFT SIDE OF A SHEET

5. Place the right hand on the counterbalance arm and the left hand on the bending leaf handle and raise the bending leaf to the position of the desired bend. *NOTE: It is sometimes necessary to raise the bending leaf a few degrees more than the required angle to allow for the resistance of the material.*
6. Return the bending leaf gradually, keeping your hand on the bending leaf handle until it is back to its former position. Open the clamping bar and remove the material.

2. Hold the material in position with the left hand and pull the clamping bar handle with the right hand until the left end of the material is firmly held. *CAUTION: Keep fingers away from the clamping bar.*
3. Place the left hand on the right end of the material and move the material until the centre of the bend line on the right end is flush with the clamping bar.
4. Finish clamping the material in position by pulling the clamping bar handle as far as possible with the right hand. (Fig. 7.11).

FIGURE 7.12   MAKING THE BEND

Assignment

1. What operations can be performed on the standard hand brake?
2. State the capacity of the hand brake in the sheet metal shop.
3. What safety precaution must be observed while using the hand brake?
4. Name the attachment that is used for forming curved shapes.
5. Why are counterbalances necessary?
6. Name the part of the hand brake that forms the metal.
7. How are the moulds held in place?
8. What is the function of the clamping bar handles?
9. Where are the bending leaf handles situated?

THE BOX AND PAN BRAKE

In principle, the box and pan brake differs little from the standard hand brake. The one significant difference is that quickly interchangeable and removable fingers allow the operator to form a box or pan, including the four sides and bottom, without damaging any one side. Thumb screws are easily loosened for the adjustment or removal of the fingers. With special attachments, tubular and radius type bends may also be formed.

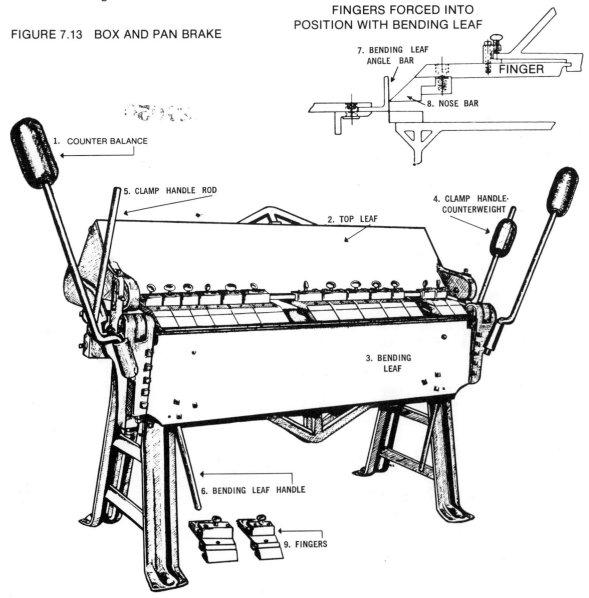

FIGURE 7.13 BOX AND PAN BRAKE

Principal Parts

1. The counterbalance offsets the weight of the bending leaf.
2. The *top leaf* is adjustable for various thicknesses of material.
3. The *bending leaf,* as the name suggests, bends the metal.
4. The *clamp handle counterbalance* counterbalances the weight of the top leaf.
5. The *clamp handle rod* raises and lowers the top leaf.
6. The *bending leaf handle* raises and lowers the bending leaf.
7. The *bending leaf angle bar* determines the bending capacity of the brake.
8. The *nose bar* determines the shape of the bend.
9. The *fingers* are designed for convenient adjustment or removal depending on the shape of the finished article.

NOTE: SAFETY PRECAUTION:

Bend only single thicknesses of material. Never bend materials heavier than the rated capacity of the brake.

Forming a Box

1. Bend the flange (1-2-3-4) to the desired angle on all four sides of the pattern (Fig. 7.14).

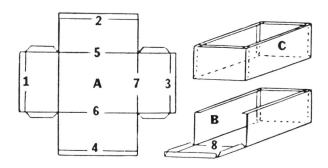

FIGURE 7.14 FORMING A BOX SHAPE WITH INSIDE FLANGES

2. Remove one or two fingers from the top leaf.
3. Adjust a suitable number of fingers to accommodate the length of the box as represented by lines 5 and 6.
4. Bend the sides of the box to 90° (Fig. 7.15).

FIGURE 7.15 FORMING THE SIDES OF A BOX

NOTE: The body of the pattern is projected to the front of the machine.

5. Readjust the fingers to accommodate the width of the box.
6. Form lines 7 and 8 to 90° (Fig. 7.16).
   CAUTION: The nose bars must be aligned evenly during the adjustment procedure.

FIGURE 7.16 FORMING THE END

Assignment

1. How does the box and pan brake differ from the standard hand brake?
2. What does the bending leaf angle bar do?
3. How many thicknesses of metal are safe to bend at once? Why?
4. What purpose do the fingers serve?
5. What sort of bends may be formed with special attachments?
6. What caution must be observed when adjusting the fingers to accommodate the width of a box?

THE SLIP-ROLL FORMING MACHINE

Roll forming machines are used to form material into curved shapes, and for this reason they are used frequently in sheet metal work. There are two general types of these machines: one has rolls mounted in solid housings and is called a roll forming machine; the other has an attachment which permits one end of the top roll to swing away from the housing and is called a slip-roll forming machine. Forming machines are equipped with three rollers; the two front rolls are gear operated, and the back roll is the idler roll.

Principal Parts

1. The *trigger release* is a small handle which is pulled forward to permit the upper roll to swing towards the operator.
2. The *operating handle* turns the two front rolls.
3. The *lower front roll* is adjusted to the thickness of the material to be formed.
4. The *housing* encloses the gears which operate the two front rolls.
5. The knurled *adjusting screws* adjust the rear and lower front rolls.
6. The *base* supports the machine.
7. The *circular grooves* at the ends of the rolls are used when rolling wire and wired edges.
8. The *upper front roll* works in conjunction with the lower front roll to feed the material to the rear roll.
9. The *release handle* is turned back to permit the upper roll to be raised by a lift handle.
10. The *rear roll* deflects the sheet upward to form a curve. It may be moved up or down to form larger or smaller cylinders.

NOTE: SAFETY PRECAUTION:

Keep fingers and loose clothing away from the rolls.

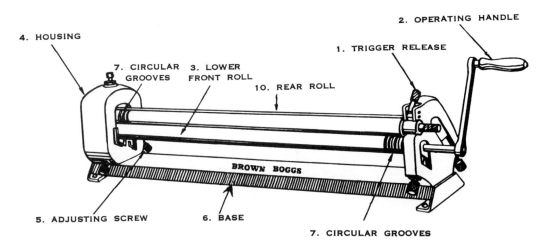

FIGURE 7.17 THE SLIP-ROLL FORMING MACHINE

Machines and Machine Operations   37

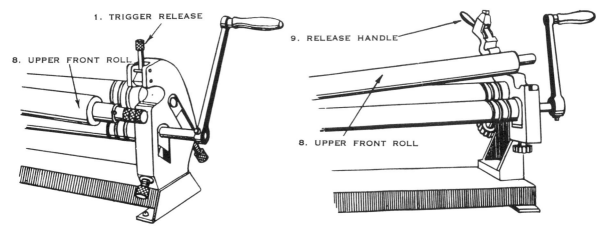

FIGURE 7.18   TYPES OF UPPER ROLL RELEASES

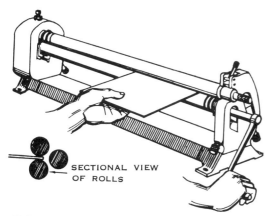

FIGURE 7.19   STARTING TO ROLL A JOB

enough, bring the sheet back to its starting position by turning the operating handle in the opposite direction, raise the rear roll slightly, and roll the sheet through the rolls. Repeat this procedure until the required curvature is obtained.

8. Release the upper roll and remove the job.

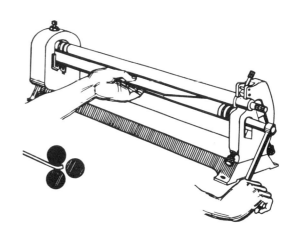

FIGURE 7.20   MAKING A STARTING BEND

Rolling Cylindrical Shapes

1. Raise or lower the bottom front roll by adjusting the two front knurled adjustment screws so that the sheet can be inserted.
2. Set the rear roll by means of the two knurled adjustment screws at the back of the machine.
3. Insert the sheet between the rolls from the front of the machine.
4. Start the sheet between the rolls by turning the operating handle (Fig. 7.19).
5. Holding the operating handle firmly with the right hand, raise the sheet with the left hand to form the starting edge. (Fig. 7.20).
6. Turn the operating handle until the sheet is partly through the rolls, changing the left hand from the front edge of the sheet to the upper edge of the sheet (Fig. 7.21).
7. Roll the remainder of the sheet through the machine. *NOTE: If the curvature is not small*

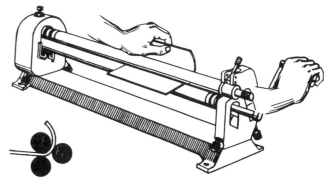

FIGURE 7.21   FORMING THE JOB

Assignment

1. What operation can be performed on the slip-roll former?
2. State the capacity of the slip-roll former.
3. What safety precaution must be observed while using the slip-roll former?
4. What advantage has a slip-roll former over an ordinary roll former?
5. Why are there grooves in the rolls?
6. Which roll is adjusted for the gauge of the metal to be formed?
7. Which roll is adjusted for the curvature required?
8. Name the part that is used to adjust the rolls.
9. What purpose does the housing serve?
10. Explain how to roll a cylinder in the slip-roll former.

THE DRILL PRESS

The drill press is used for several operations, the most common of which is drilling holes. There are many types of drill presses, but they all work on the same principle: a twist drill is rotated by electric power. All drill presses have some means of holding the work on the table and adjusting the speed of the spindle for different sizes of drills and kinds of materials. The capacity of a drill press is expressed as the maximum diameter of stock that can be drilled through the centre on the machine.

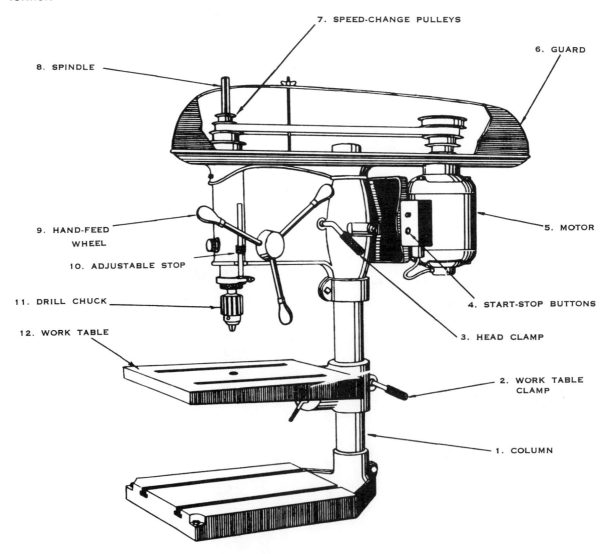

FIGURE 7.22   THE DRILL PRESS

Principal Parts

1. The *column* supports the work table and the head.
2. The *work table clamp* holds the work table in place.
3. The *head clamp* holds the head in place.
4. The *start-stop* buttons.
5. The *motor* supplies power to drive the spindle.
6. The *guard* encases the pulleys and the belt.
7. The *speed-change* pulleys regulate the spindle speed.
8. The *spindle* turns the chuck and the drill.
9. The *hand-feed wheel* raises and lowers the drill.
10. The *adjustable stop* limits the depth of the hole to be drilled.
11. The *drill chuck* holds the drill and turns it.
12. The *work table* supports the work.

NOTE: SAFETY PRECAUTION:

Never make adjustments on the machine while it is running.

Drilling Small Holes

1. Centre punch the holes to be drilled.
2. Clamp the work and a wood block to the work table.
3. Select the proper size of drill by using a drill table and reading the size at the shank.

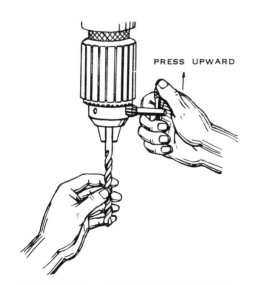

FIGURE 7.23 CHUCKING THE DRILL

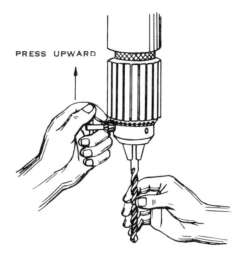

FIGURE 7.24 REMOVING THE DRILL

4. Insert the drill in the chuck and tighten the jaws of the chuck with the key. Fig. 7.23 shows the proper method of chucking a drill, and Fig. 7.24 shows how to remove a drill.
5. Start the machine and feed the drill slowly until the point is started in the metal. Raise the drill and check to see if the drill is in the proper location. If it is not, relocate the hole with a centre punch.
6. Lower the drill to the work and feed it to the required depth.
7. Remove and clean the drill. Return it to the proper rack.
8. Remove the work and file off the burrs. Clean the work table.

Assignment

1. What operation is performed on the drill press?
2. Why is it necessary to adjust the speed of the spindle?
3. What safety precaution must be observed while using the drill press?
4. Why is it necessary to encase the pulleys and belt?
5. How are small holes drilled?
6. What is the purpose of a chuck key?
7. How are small parts held when they are being drilled?
8. Make a neat sketch of chucking the drill.
9. How do you locate the centres of holes to be drilled?
10. What is the function of the work table?

THE BAR AND TUBE BENDER

The bar and tube bender is a tool mounted, manually operated, forming machine. The mounted tool consists of a form or radius collar having the same shape as the desired bend, a clamping block or locking pin that securely grips the material during the bending operation, and a forming roller or follow block which moves around the bending form.

Since all metals are somewhat elastic they will spring back after they are formed. The amount of spring back is dependent upon the type of material, its size and hardness, and the radius of the bend. Therefore, it is usually necessary to experiment before determining exact bending form sizes. Generally the results obtained will be in proportion to the care taken in properly tooling the bender for the job to be done. Attachments and parts are available to do various forming operations with bars, wire, and tubes, as long as they are within the capacity of the machine as rated by the manufacturer.

Principal Parts

1. The *operating arm* is used to exert sufficient pressure on the material being formed.
2. The *forming nose* leads the material being formed around the contour setting.
3. The *radius pin* determines the diameter for a desired circular bend.
4. The *locking pin* provides a positive means of clamping the material against the contour forms.
5. The *angle gauge* is adjusted for the required degree of bend.
6. The *bend locating gauge* is adjusted to locate material at required lengths from the bend position.
7. The *return stop* limits the opening of the forming nose.
8. The *accessory mounting bolt* is used as a clamping device for special forming arrangements.

NOTE: SAFETY PRECAUTION:

The bender should be mounted in a location which will allow the operator freedom of operation around the entire machine.

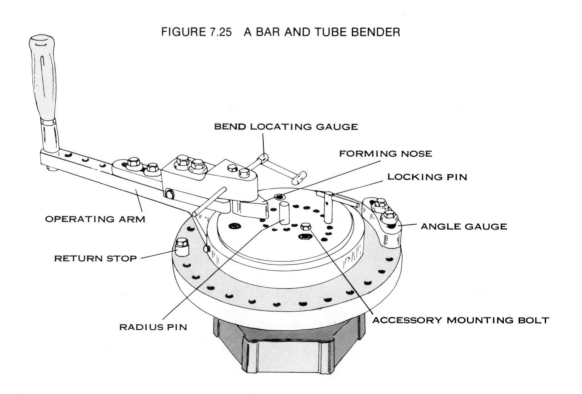

FIGURE 7.25 A BAR AND TUBE BENDER

Scroll Bending

As the material will only bend where the contour collar offers resistance, the forming nose can lead the material around until it contacts the "highpoint" and exerts sufficient pressure to force it into the shape of the collar.

Square Bending

Forming zero radius bends around square, rectangular or other multi-sided blocks employs the same principle used in scroll bending.

FIGURE 7.26

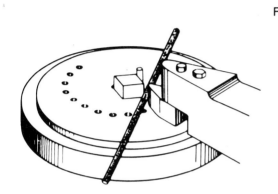

Adjust Forming Nose so material will fit snugly between Nose and any edge of Square Block.

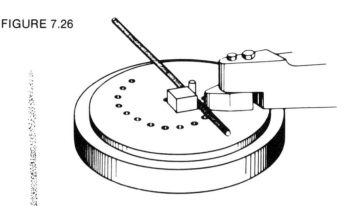

Clamp material between Locking Pin and Square Block and then advance Operating Arm.

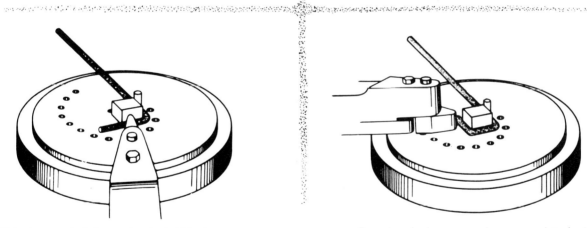

Note how material remains straight between corners of block as Forming Nose moves into position for second bend.

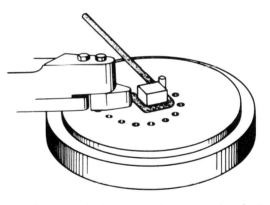

Two bends have now been completed. A third bend can be obtained by merely advancing the Operating Arm.

Assignment

1. What determines the amount of spring back of a metal?
2. How can good results be obtained from the bar and tube bender?
3. What does the bend locating gauge do?
4. What is the purpose of the forming nose?
5. Sketch a bar and tube bender.
6. What is the purpose of the return stop?

42   Sheet Metal Practice

THE ROD PARTER

The rod parter is used to accurately cut or "part-off" rods, wire and bars without distorting or crushing the material. Although most shearable materials can be accurately cut without leaving a burr, the squareness of the cut is largely determined by the quality of the material, as "parting-off" is a combination shearing-breaking operation.

Principal Parts

1. The *handle* activates the cutting dies.
2. The *handle link* provides a multiple lever arrangement, making cutting a relatively easy operation.
3. The *cutting dies* are designed to cut the material. They are arranged around a central axis and come in graduated sizes to match the diameter of the material being cut.
4. The *pivot bolt* is used to align the cutting dies.
5. The *upper casting* supports the cutting dies.
6. The *base casting* is fastened to a suitable support base, such as a bench or table.
7. The *gauge* can be adjusted for repeat cutting of short lengths of stock.

NOTE: SAFETY PRECAUTION:

Never exceed the maximum cutting capacity of the cutting dies.

Gauge Operation

1. Adjust the gauge for the desired diameter and length of stock. Only the end portion of the material should contact the gauge.

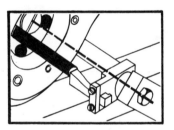

FIGURE 7.28A   ADJUSTING THE GAUGE

2. When the material is stopped by the gauge, pull the handle. This will cut the material at the desired length.

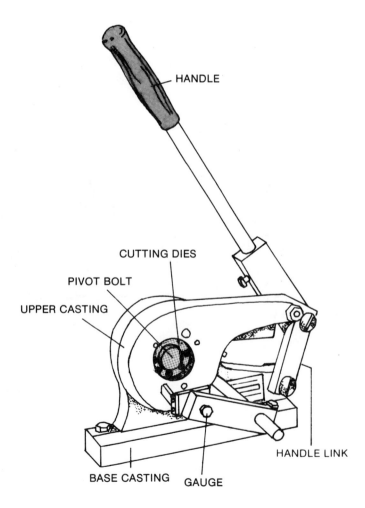

FIGURE 7.27   A ROD PARTER

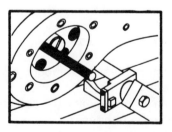

FIGURE 7.28B   CUTTING THE MATERIAL

3. Return the handle to its former position.

NOTE: The cut piece is automatically ejected and the next piece is gauged for cutting.

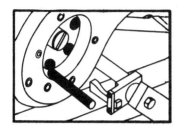

FIGURE 7.28C   THE FINISHED CUT

Assignment

1. What advantage does the rod parter have over other methods of cutting rods, wires, and bars?
2. What safety precaution must be observed when using the rod parter?
3. Explain the function of the gauge.
4. What portion of the material should contact the gauge?
5. What determines the squareness of the cut which can be achieved with the rod parter?
6. What purpose does the handle link fulfill?

THE LOCKFORMER

Lockformer-made Pittsburgh locks and other connections assemble faster and easier than work done on brakes and folders. Lockformer Pittsburgh locks, for example, are always uniformly open. Standing edges or flanges press easily into their recesses.

Besides Pittsburgh locks lockformers can also make double seam locks and drive cleats and right angle flanges on straight pieces. If a power flanging attachment is added right angles can be formed on either straight or curved pieces.

Lockformers handle aluminum and other metals in addition to conventional galvanized iron. Material of any length over 180 mm may be formed without complicated adjustment, when forming varying gauges of material.

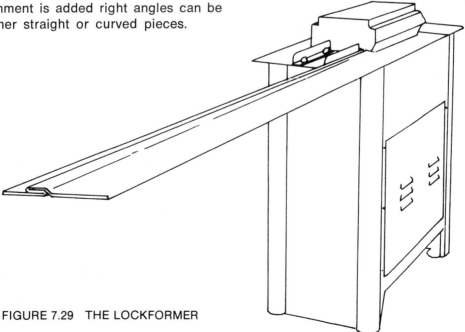

FIGURE 7.29   THE LOCKFORMER

FIGURE 7.30 PRINCIPAL PARTS OF A LOCKFORMER

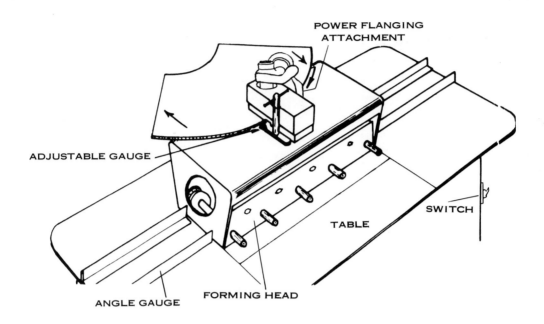

Principal Parts

1. The *table* supports the work being formed.
2. The *angle gauge* is adjustable to the desired width for the pockets of the Pittsburgh lock.
3. The *forming head* contains the forming rolls that determine the shape of the finished Pittsburgh lock.
4. The *power flanging attachment* is adjustable to turn up flanges at 90°.
5. The *switch* is used to start and stop the lockformer.

NOTE: SAFETY PRECAUTION:

Material less than 180 mm in length should never be fed into the lockformer.

Maintenance

Check and refill the grease cups on the underside of the forming head at least once a month. The slow speed shafts do not require lubrication. Check the oil in the motor occasionally and add as required.

Forming a Pittsburgh Lock

1. Switch the power source on.
2. Place the raw edge of the material against the angle gauge.
3. Insert the leading edge of the work to be formed into the lockformer rolls.
4. Remove the work from the table after it has been machine formed.
5. Turn off switch.

Operating the Power Flanging Attachment

1. Turn the front adjusting screw all the way in and then back ⅛ to ¼ of a turn.
2. Tighten the adjusting dial on the back side of the flange to the stop and turn the dial back to the proper gauge setting.
3. Turn up a starting flange at approximately 45° in the slot provided.
4. Insert the starting flange into the forming head.

NOTE: As the material passes through the rolls, the compensator arm will make contact with the material and guide it through the rolls.

FIGURE 7.31 THE POWER FLANGING ATTACHMENT

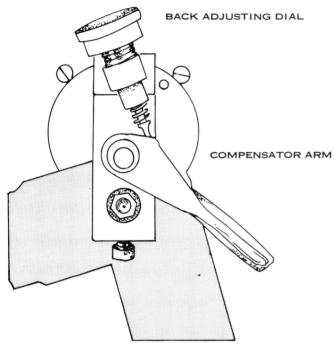

Standard Auxiliary Rolls	
⌐⌐	Pittsburgh Rolls
⌐⌐	Double Seam Rolls
⌐⌐	Drive Cleat Rolls
⌐⌐	Flange Rolls

Assignment

1. What advantage do lockformer made locks have over work done on brakes and folders?
2. What can lockformers do besides making Pittsburgh locks?
3. What kind of metals can be used in a lockformer?
4. What is the purpose of the angle gauge?
5. What part or parts determine the shape of a finished Pittsburgh lock?
6. State the capacity of the lockformer.
7. Why should material shorter than 180 mm not be fed into the lockformer?
8. Name four standard auxiliary rolls.
9. Describe how to operate the power flanging attachment.

THE TURNING MACHINE

The turning machine is a rotary machine. It is used for forming a narrow edge with a small radius on circular jobs. If the edge is made to project at 90° or beyond it may be used for double seaming, wiring, or stiffening an edge.

Principal Parts

1. The *operating handle* turns the upper and lower gear-connected rolls.
2. The *upper roll* forms the edge of the work by pressing the metal into the groove on the lower roll.
3. The *lower roll* is aligned to accommodate the upper roll.
4. The *crankscrew* moves the upper roll.
5. The *knurled adjusting screw* moves the gauge to the required setting.
6. The *knurled lock nut* holds the gauge in place after the setting has been determined.
7. The *gauge* face provides a stable surface for the raw edge of the work.
8. The *bench standard* is clamped to a suitable surface.

NOTE: SAFETY PRECAUTION:

Fingers and loose clothing should be kept clear of the rolls at all times.

FIGURE 7.32 THE TURNING MACHINE

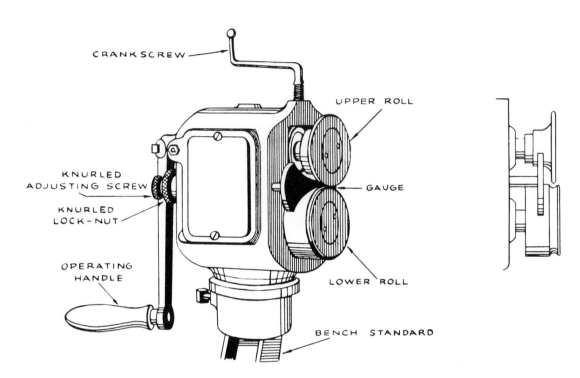

Turning an Edge for Wiring

1. Centre the upper and lower rolls.

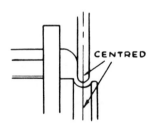

FIGURE 7.33 CENTRING THE ROLLS

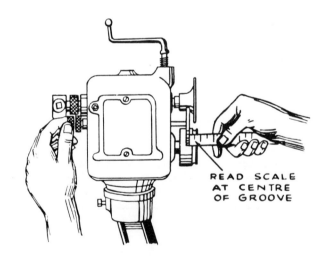

FIGURE 7.34 SETTING THE GAUGE

2. Set the gauge for the width of edge required by using the knurled screw at the end of the gauge. Use a ruler to measure from the gauge to the centre of the edge of the upper roll.

NOTE: *For a wired edge the width of material to be turned must be equal to 2.5 times the diameter of the wire.*

3. Using the knurled lock-nut, lock the gauge in place.

Machines and Machine Operations 47

4. Holding the workpiece in the left hand, position the edge to be turned between the rolls and against the gauge.
5. Lower the upper roll by turning the crankscrew with the right hand until it just touches the metal, making a slight impression.

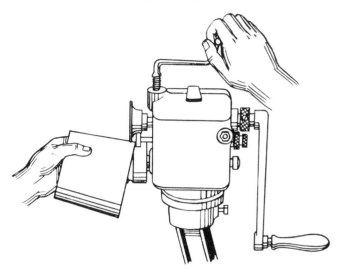

FIGURE 7.35 UPPER ROLL ADJUSTMENT

6. With the left hand hold the edge of the workpiece against the gauge, then turn the operating handle with the right hand to make one complete turn on the workpiece.
7. Turn the crankscrew to lower the upper roll slightly.
8. Raise the workpiece slightly and use the operating handle to make another complete turn.

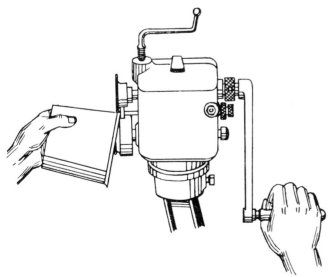

FIGURE 7.36 STARTING THE EDGE

9. Again lower the upper roll. Complete the edge by turning the operating handle and raising the workpiece gradually after each revolution until the required edge is obtained.

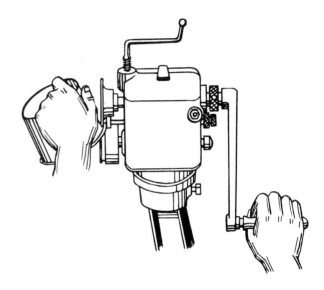

FIGURE 7.37 COMPLETING THE EDGE

10. Loosen the crankscrew and remove the workpiece.

Assignment

1. What is the main function of the turning machine?
2. To what other uses can the turning machine be put?
3. Describe the function of the upper and lower rolls.
4. Describe how you would set the gauge for the width of edge required.
5. If a wired edge is desired how much of the workpiece must be allowed for turning?
6. Why is it wise to keep fingers and loose clothing away from the rolls?
7. What part moves the gauge to the required setting?

THE BURRING MACHINE

Like the turning machine the burring machine is also a rotary machine. It is used to form narrow edges on covers, cylinders, discs, and irregular pieces. These narrow edges are called burred edges. The process of making such an edge is called burring. Double seams, set-in bottoms and other types of corner seaming are typical uses for burred edges.

Principal Parts

1. The *operating handle* turns the upper and lower gear-connected rolls.
2. The *upper roll* clamps the metal while the edge is being burred.
3. The *lower roll* is recessed near the outer edge to provide a sharply radiused burr.
4. The *crankscrew* adjusts the upper roll up and down.
5. The *knurled adjusting screw* moves the gauge to the required setting.
6. The *knurled lock-nut* holds the gauge in place after the setting has been determined.
7. The *gauge* face provides a stable surface for the raw edge of the work.
8. The *bench standard* is clamped to a suitable surface.

NOTE: SAFETY PRECAUTION:

A hand guard made of a small piece of metal bent in a U-shape must be used when operating the burring machine. The guard serves to protect the hand and also allows the job to turn freely.

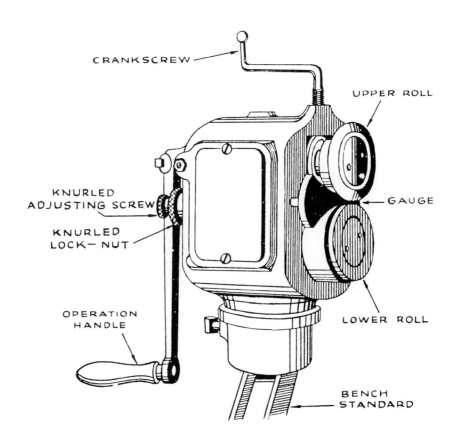

FIGURE 7.38   THE BURRING MACHINE

Burring a Disc

1. Space the rolls as indicated in Figure 7.39. The knurled adjusting screw at the opposite end of the machine controls the spacing.

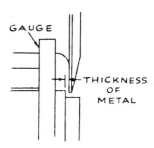

FIGURE 7.39 HOW TO SPACE THE ROLLS

2. Place one end of a rule against the gauge as indicated in Figure 7.40. The distance between the gauge and the inner edge of the upper roll must be equal to the width of the burr required. Turn the knurled adjusting screw until the proper distance can be read on the rule.

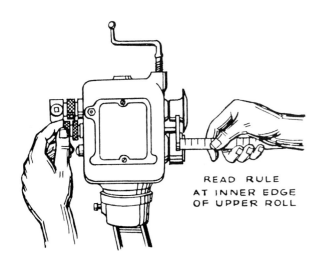

FIGURE 7.40 ADJUSTING AND SETTING THE GAUGE

3. The gauge is then locked in place by means of the knurled lock-nut.
4. With the hand guard in place hold the disc in the left hand and place the edge to be burred between the rolls and against the gauge.

5. Turn the crankscrew with the right hand, bringing down the upper roll until the metal disc is firmly held between the rolls.

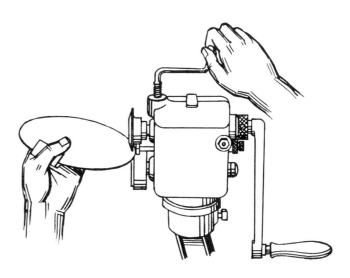

FIGURE 7.41 PREPARING FOR THE BURR

6. Start the burr by holding the disc against the part of the gauge between the rolls. Turn the operating handle to rotate the work and start the burr. Allow the disc to pass freely through the guard.

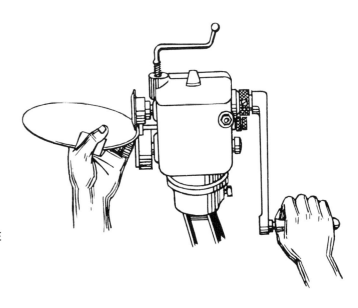

FIGURE 7.42 STARTING THE BURR

7. Burr the disc around its entire circumference, raise it slightly, and repeat the procedure. Continue this process until the edge is burred to the position required.

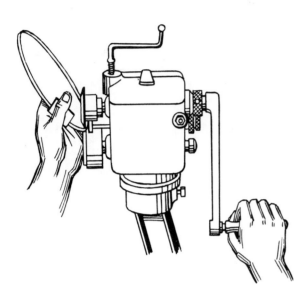

FIGURE 7.43   FINISHING THE BURR

NOTE: *Small discs which have buckled may be straightened by throwing them on a flat surface, burred edge up.*

Assignment

1. Name five parts of the burring machine.
2. What sort of shapes can be burred on the burring machine?
3. Name two uses for burred edges.
4. What special safety device must be used when burring edges?
5. Describe the first three steps in burring a disc.
6. How may small discs which have buckled be straightened?
7. What is the function of the upper roll?
8. What must a disc be held against when starting a burr?

THE EASY EDGER

The easy edger turns a uniform right angle 5 mm edge on any radius over 75 mm and on any curve or irregular shape in a single operation. Unlike other rotary machines the easy edger may be operated from the right or left hand side. This specially designed machine is one of the most useful pieces of equipment to be found in any ventilating or air conditioning sheet metal shop.

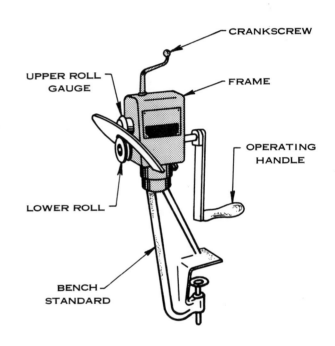

FIGURE 7.44   THE EASY EDGER

Principal Parts

1. The *operating handle* turns the upper and lower gear-connected rolls.
2. The *upper roll* forms the edge of the work by pressing the metal into the shoulder of the lower roll.
3. The *lower roll* is aligned to accommodate the upper roll.
4. The *crankscrew* raises and lowers the upper roll.
5. The *guide-plate* is mounted at a fixed 90° angle to the inside edge of the lower roll. The face of the work is positioned on the guide-plate.
6. The *bench standard* is clamped to a suitable surface.

NOTE: SAFETY PRECAUTION:

Keep fingers and loose clothing away from the rolls.

Turning a Right Angle Edge

1. Crease or score the work to be turned 5 mm in from the edge.
2. Place the material to be edged between the rolls as indicated. The edge of the material must rest securely in the 90° shoulder of the bottom roll.

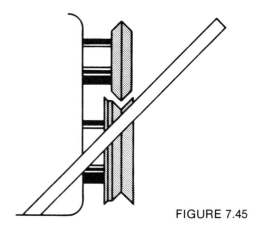

FIGURE 7.45

3. Lower the upper roll by turning the crankscrew until the work is firmly held between the rolls. Do not turn the crankscrew too tight as this would damage the work.
4. Grasp the work with one hand and turn the operating handle with the other hand.

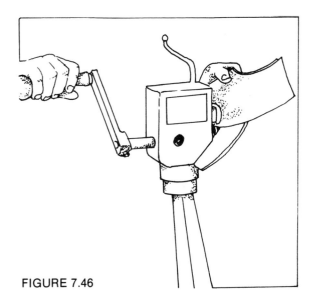

FIGURE 7.46

NOTE: The material must be held securely against the shoulder on the bottom roll as the operating handle is turned.

5. If after one complete pass through the rolls the turned edge is buckled or under 90°, tighten the crankscrew and repeat step #4.

Assignment

1. What happens if the crankscrew is turned too tight?
2. What is the minimum radius the easy edger can handle?
3. How does the easy edger differ from other rotary machines?
4. What should you do if a turned edge comes out at less than 90°?
5. What sort of machine is the easy edger? Name two other machines which fit in this category.
6. What is the first step in turning a right angle edge?
7. Make a sketch of the upper and lower rolls of the turning machine and the easy edger. How are they different?

THE TURRET PUNCH PRESS

The turret punch press consists of two cylindrical heads or turrets mounted on the upper and lower arms of a C-shaped frame. Punches are located on the upper turret and dies on the lower one. Both turrets may be rotated in either direction until the desired punch and die set is selected. Accurate alignment of the set is then assured by the use of safety locks.

Particularly because of the great variety of punch and die sizes and shapes available the turret punch press is very useful for any punching program requiring round or irregular holes, whether punched singly or in groups with the use of special gauges.

Principal Parts

1. The *frame* supports the upper and lower cylindrical turret heads.
2. The *throat* capacity is the space between the frame in back of the turret. Its depth permits the operator to punch holes in the centre of large wide work.
3. The *punches and dies* are prealigned for proper punching over a wide dimensional range.
4. The *quick turn handles* enable the operator to bring the desired punch or die into operating position.
5. The *alphabet indicators* clearly mark the punches and dies to eliminate the possibility of misalignment.
6. The *safety locks* prevent damage to the work, punches, and dies.
7. The *handle,* when pulled, operates the punches and dies.
8. The *floor stand* provides a suitable mount for the frame.

NOTE: SAFETY PRECAUTION:

Special clearance allowances must be observed when heavy gauge materials are readied for punching.

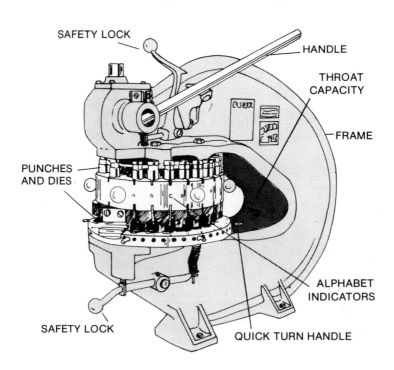

FIGURE 7.47   THE TURRET PUNCH PRESS

Maintenance

The turret punch must be oiled periodically, especially when it is used frequently.

Punching Procedure

1. Release the safety lock on the punch turret.
2. Select the desired punch and swing the punch turret into the punching position. Return the safety lock.
3. Release the safety lock on the die turret.
4. Select the desired die and swing the die turret into the punching position. Return the safety lock.
5. Place the centring point of the punch on the centre mark of the work to be punched.

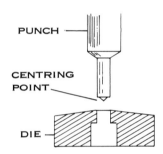

FIGURE 7.48 PUNCH AND DIE

6. Punch the work by pulling the handle forward.
7. Return the handle to its former position on the handle rest.

Assignment

1. What precaution should be taken before punching heavy gauge materials?
2. What maintenance does the turret punch press require?
3. What is throat capacity?
4. What ensures proper alignment of the punches and dies?
5. What direction must the turrets be rotated?
6. Describe the punching procedure step by step.

THE SPOT WELDER

Spot welding is a welding process in which the heat necessary to fuse material together is generated by resistance to a flow of electrical current. Spot welding is accordingly a type of resistance welding. The work under assembly resists the flow of current and is heated to a plastic state, causing it to bond together. Because heat is created, spot welders need an adequate cooling system. Controls are required to regulate electrical resistance. Welding time is controlled by an automatic electronic timer on some machines. On other spot welders the welding time is controlled totally by means of the foot pedal.

NOTE: Materials to be spot welded must be clean and free of grease or oil if good strong welds are desired.

Principal Parts

1. The *foot pedal* is used to operate the spot welder.
2. The *trip switch* controls the flow of electrical energy through the tips.
3. The *electrode tips* contact and compress the workpieces and provide the heat for welding. The tips must be kept cooled by a water circulation system so they do not overheat.
4. The *rocker arm* allows the distance between the tips to be adjusted for a wide range of material thicknesses.
5. The *voltage regulator* controls the electrical current.
6. The *electronic timer* may be adjusted to control the welding time. Not all spot welders are equipped with electronic timers, and the welding time is not adjustable on these machines.
7. The *electrodes* conduct the flow of electrical energy on circuit when welding heat is required.
8. The *water* connections are connected to a water supply. Water flows from the connections to the electrode tips and back.

NOTE: SAFETY PRECAUTION:

Keep the machine clean. Any corrosion of copper or bronze parts will result in deterioration in the conductive capacity of the machine.

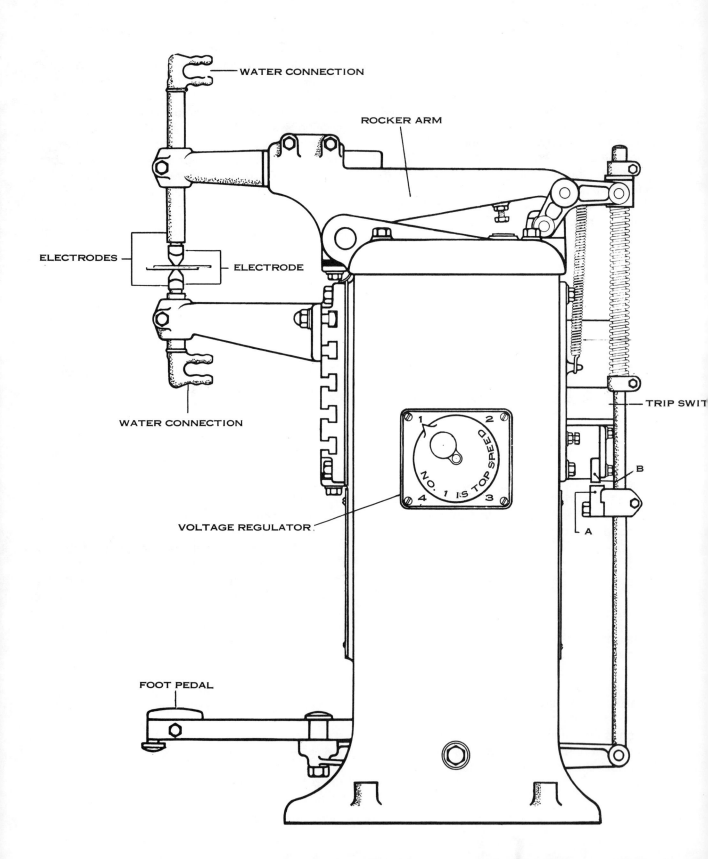

FIGURE 7.49 THE SPOT WELDER

Operating Procedure

1. Turn on the electric power source and the water supply.
2. Adjust the voltage regulator and electronic timer settings.

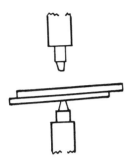

FIGURE 7.50

3. Insert the properly aligned workpieces between the open electrode tips.

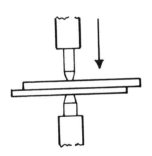

FIGURE 7.51

4. Apply pressure on the foot pedal, causing the tips to contact the work pieces.
5. Release the foot pedal when it touches bottom.

NOTE: On spot welders with electronic timers the weld is completed at the end of the timing cycle. The foot pedal should not be released until the timer automatically interrupts the energy supply.

Assignment

1. What is spot welding?
2. What kind of energy is used in spot welding?
3. Why must the electrode tips be water cooled?
4. What is the foot pedal used for?
5. Why should spot welders be kept clean?
6. What is the purpose of the rocker arm?
7. When should the foot pedal not be released on spot welders equipped with electronic timers?
8. Describe what happens to the material when a spot weld is made.

PART TWO
Projects

Rectangular Box 1	58
Rectangular Box 2	60
Cookie Sheet	61
Spike File	63
Spice Rack	65
Utility Box	67
Waste Basket 1	70
Waste Basket 2	72
Levered Handle Dustpan	74
Fixed Handle Dustpan	77
Camp Cup	80
Hanging Planter	82
Windowsill Planter	84
Clothes Peg Box	86
Corner Shelf	90
Three Ring Book Binder	92
Coal Scuttle Ash Tray	94
Aquarium	96
Fishing Tackle Box	98
Watering Can	101
Shoeshine Box	104
Filing Cabinet	107
Filing Tray	110
Carry All	113
Tool Box	115
Barbecue	118
Picnic Cooler	124
Funnel	131

NOTE: All project drawings are dimensioned in millimetres.

RECTANGULAR BOX 1

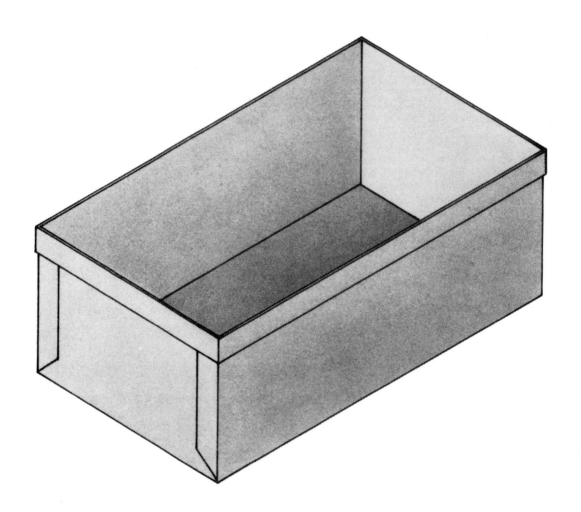

Bill of Material

Item	**No. Req'd.**	**Material**	**Pre-Cut Size**
Box	1	IC tin plate	200 mm x 150 mm

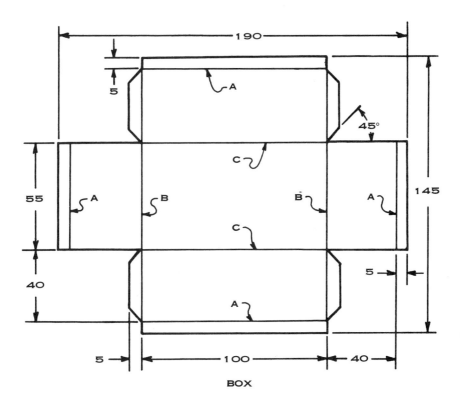

BOX

Instructions

1. Cut the metal to the required size.
2. Lay out as shown.
3. Cut the pattern to shape.
4. Make closed folds on lines marked "A", away from the lines.
5. Make square bends on lines marked "B", towards the lines. Repeat on lines marked "C".
6. Complete forming with a mallet on a suitable stake.

    NOTE: The tops of laps are inserted under the single hem.

7. Solder the corner lap joints neatly.
8. Dress all sharp edges.

RECTANGULAR BOX 2

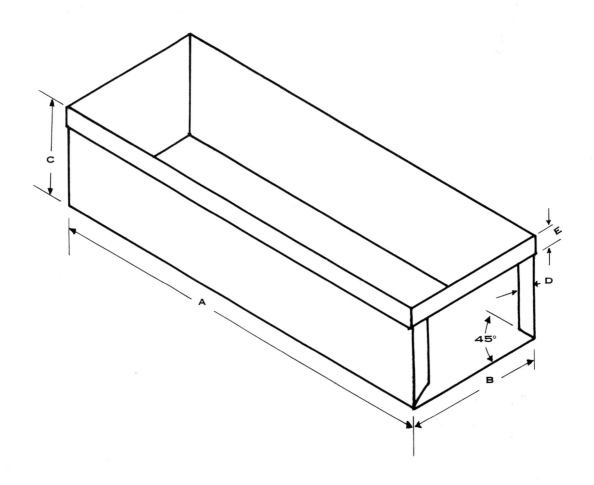

Bill of Material

Item	No. Req'd.	Material	Pre-Cut Size
Box	1	as req'd.	as req'd.

MATERIAL	A	B	C	D	E

Instructions

1. Draw the above chart in your notebook. Fill in the sizes required.
2. Draw a profile view to determine cutting size.
3. Lay out and notch the pattern.
4. Form, solder, or spot weld as required.
5. Dress all sharp edges.

COOKIE SHEET

Bill of Material

Item	No. Req'd.	Material	Pre-Cut Size
Sheet	1	IX tin plate	400 mm x 325 mm

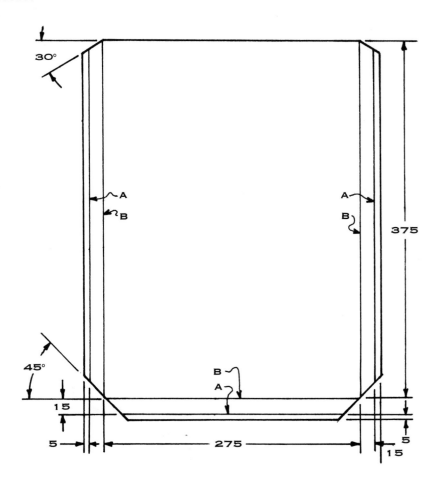

Instructions

1. Cut the metal to the required size.
2. Lay out as shown.
3. Cut the pattern to shape.
4. Make closed folds on lines marked "A", away from the lines.
5. Make square folds on lines marked "B", towards the lines.
6. Dress all sharp edges.

SPIKE FILE

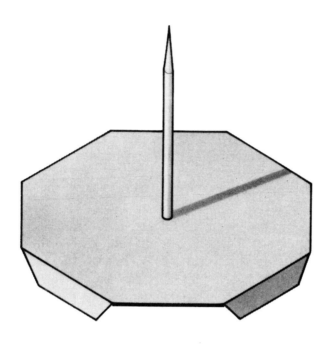

Bill of Material

Item	No. Req'd.	Material	Pre-Cut Size
Base	1	24 Ga. Satin Coat	150 mm x 150 mm
Spike	1	#10 coated wire	100 mm long

64 Sheet Metal Practice

Instructions

1. Cut the metal to the required size.
2. Lay out as shown.
3. Cut the pattern to shape.
4. Drill or punch the 3 mm diameter hole.
5. Make closed folds on lines marked "A".
6. Make square bends on lines marked "B".
7. Repeat on lines marked "C".
8. Grind the wire to a point at one end.
9. Insert the plain end through the hole about 3 mm.
10. Solder in place.
11. Dress all sharp edges.

SPICE RACK

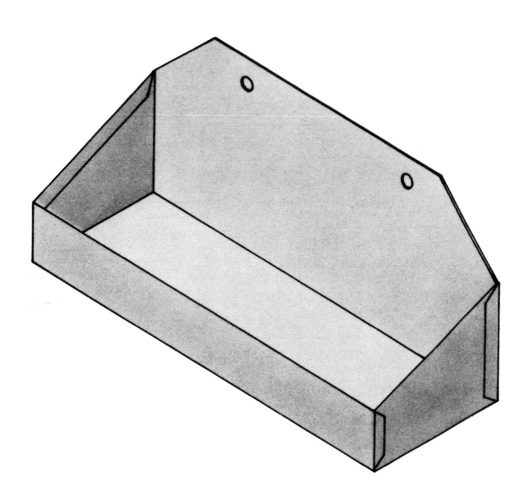

Bill of Material

Item	No. Req'd.	Material	Pre-Cut Size
Rack	1	IX tin plate	350 mm x 250 mm

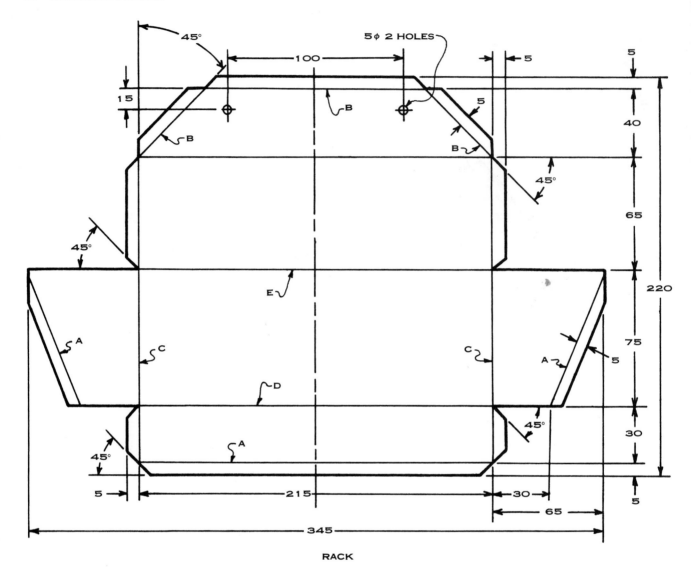

RACK

Instructions

1. Cut the metal to the required size.
2. Lay out as shown.
3. Cut and notch the pattern to shape.
4. Make closed folds on lines marked "A", towards the lines.
5. Make closed folds on lines marked "B", away from the lines.
6. Make square bends on lines marked "C", towards the lines.
7. Make a square bend on line "D", towards the lines. *NOTE: The soldering laps are to be placed outside.*
8. Repeat step 7 for line "E".
9. Solder the seams neatly.
10. Remove all sharp edges.

UTILITY BOX

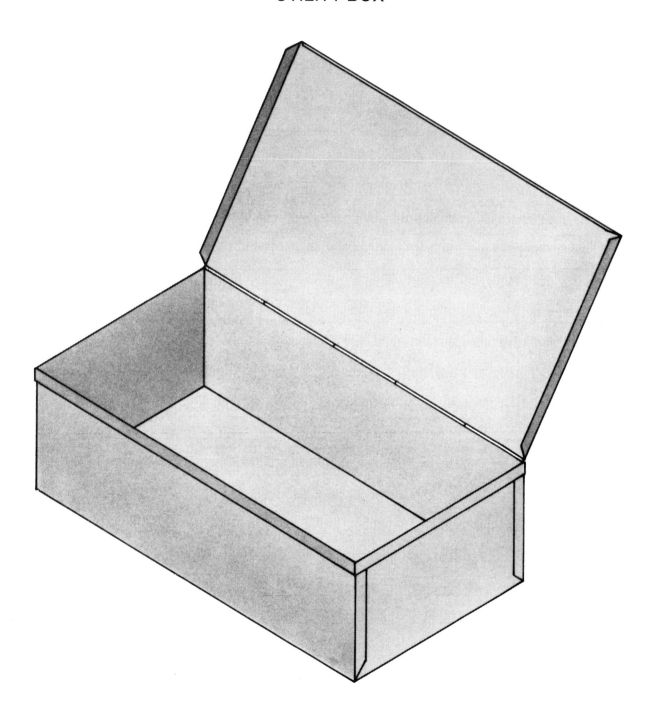

Bill of Material

Item	No. Req'd.	Material	Pre-Cut Size
Box	1	IX tin plate	350 mm x 250 mm
Cover	1	IX tin plate	250 mm x 150 mm
Hinge			
Pin	1	#10 wire	205 mm long

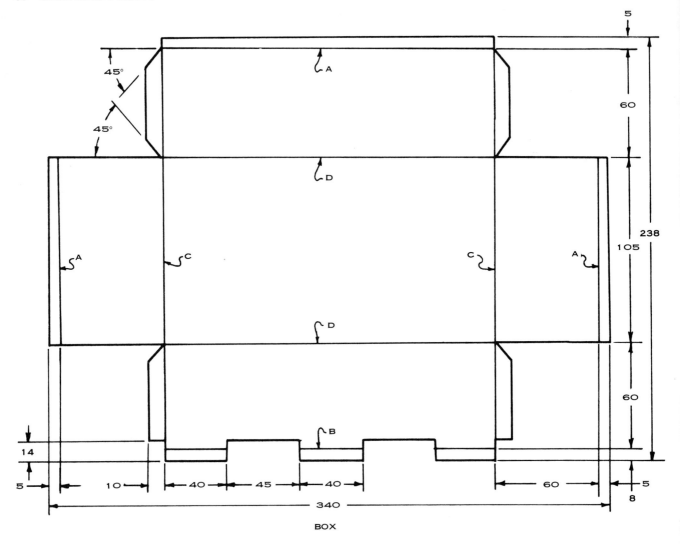

BOX

Instructions — Box

1. Cut the material to the required size.
2. Lay out as shown.
3. Cut the pattern to shape.
4. Make closed folds on the lines marked "A", away from the lines.
5. Make a square fold on line "B", towards the line.
6. Obtain a piece of #10 wire, 205 mm long, and fasten in fold "B".
7. Make square bends on the lines marked "C", letting the wired edge go past the end of the clamping bar.
8. Make square bends on the lines marked "D".
9. Complete forming with a mallet on a suitable stake.
10. Solder the laps carefully.
11. Dress all sharp edges.

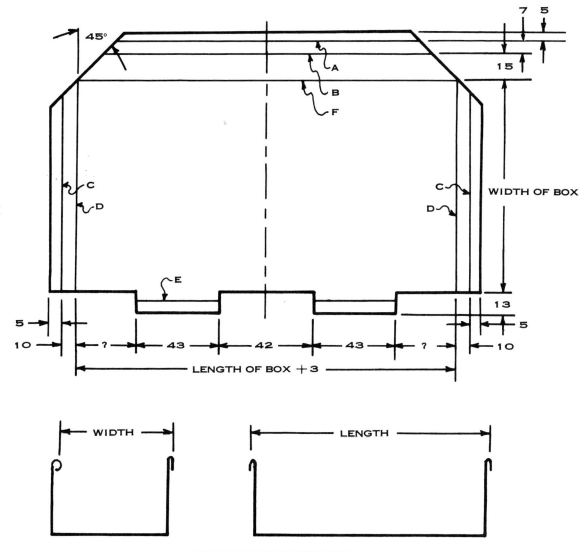

FORMED PROFILE OF BOX

Instructions — Cover

1. Determine the exact cutting size of the cover layout.
2. Cut the material to the required size.
3. Lay out and notch the pattern.
4. Make closed folds on the lines marked "A", "B", and "C", towards the lines.
5. Make square folds on the lines marked "D", "E", and "F", towards the lines.
6. Place the tabs for the hinge under the wire on the box, and bend them around wire.
7. Check to see that cover fits and operates satisfactorily.
8. Dress all sharp edges.

WASTE BASKET 1

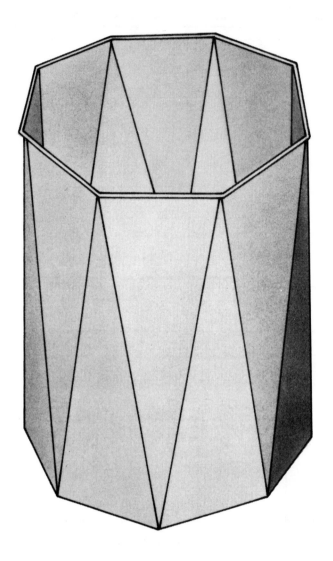

Bill of Material

Item	No. Req'd.	Material	Pre-Cut Size
Body	1	IC tin plate	700 mm x 250 mm
Bottom	1	IX tin plate	225 mm x 225 mm
Wire	1	#12 coated	700 mm long

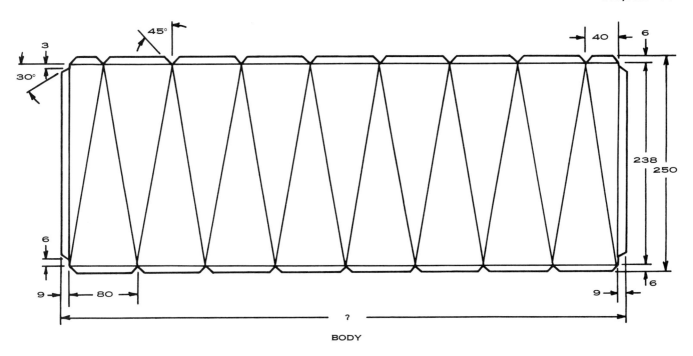

BODY

Instructions — Body and Bottom

1. Cut the material to the required size.
2. Lay out the body as shown.
3. Notch the pattern to shape.
4. Make 6 mm full folds on each end. The folds are to face opposite sides.
5. Make a 6 mm full fold along the bottom edge, away from the lines.
6. Make a 6 mm square fold along the top edge, towards the lines.
7. Lay out and notch the bottom as shown.
8. Make 6 mm square folds on all sides, towards the lines.
9. Complete forming of oblique lines on the body on a suitable stake, away from the lines. Use the bottom as a guide.
10. Complete the grooved seam.
11. Obtain a piece of #12 wire and wire the top edge.
12. Place the bottom in position, close all folds, and tack solder at the points.
13. Dress all sharp edges.

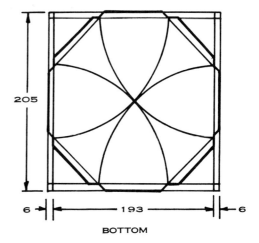

BOTTOM

72 Sheet Metal Practice

WASTE BASKET 2

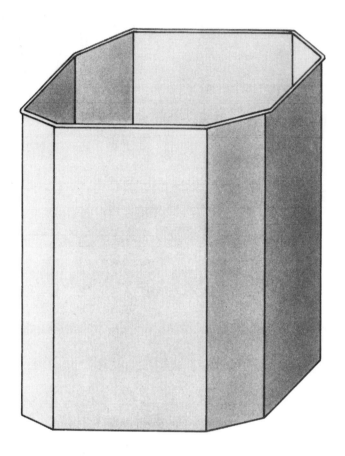

Bill of Material

Item	No. Req'd.	Material	Pre-Cut Size
Body	1	IC tin plate	700 mm x 250 mm
Bottom	1	IX tin plate	225 mm x 225 mm
Wire	1	#12 coated	700 mm long

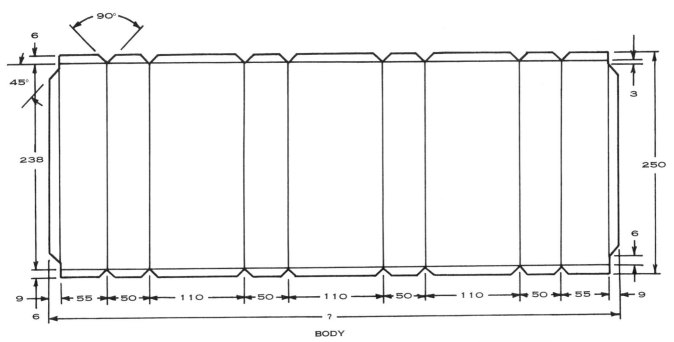

BODY

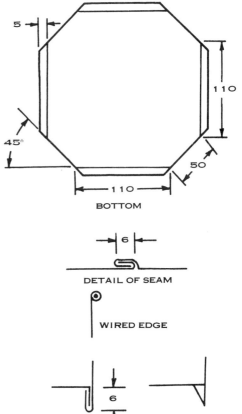

Instructions — Body

1. Cut the material to the required size.
2. Lay out as shown.
3. Notch the pattern to shape.
4. Make 6 mm full folds on each end. The folds are to face opposite sides.
5. Make a 6 mm full fold along the bottom edge, towards the lines.
6. Make a 6 mm square fold along the top edge, away from the lines.
7. Make 45° bends on vertical lines in the brake or on a suitable stake.
8. Complete the grooved seam.
9. Complete wiring of the top edge.

Instructions — Bottom

1. Lay out and notch as shown.
2. Make 5 mm square folds on the long sides.
3. Place the bottom in as per detail and solder.
4. Remove all sharp edges.

LEVERED HANDLE DUST PAN

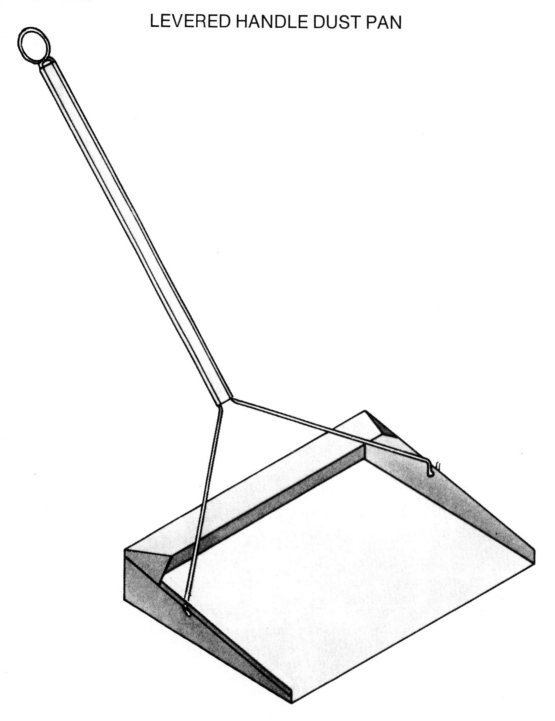

Bill of Material

Item	No. Req'd.	Material	Pre-Cut Size
Pan	1	26 Ga. galv. iron	350 mm x 350 mm
Rest	1	26 Ga. galv. iron	225 mm x 40 mm
Handle	1	26 Ga. galv. iron	450 mm x 50 mm
Wire	1	#10 coated	1600 mm long

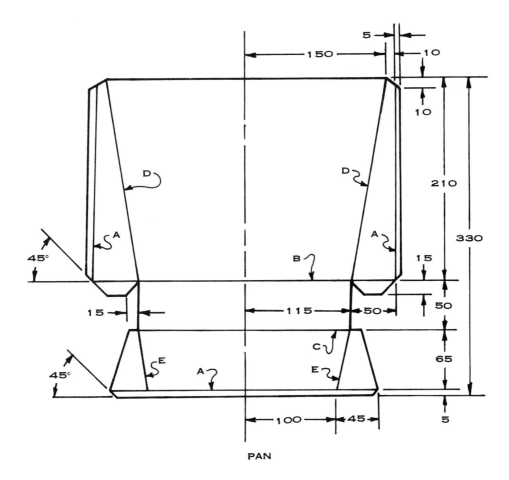

PAN

Instructions — Pan

1. Lay out as shown. *NOTE: All vertical lines and oblique lines are measured from the centre line.*
2. Cut and notch the pattern to shape.
3. Make closed folds on the lines marked "A", away from the lines.
4. Make a square bend on line "B", towards the line.
5. Bend line "C" in the same direction until the bottom touches the clamping bar.
6. Make square bends on the lines marked "D".
7. Solder the laps in place.
8. Bend over a stake on the lines marked "E" until the hood meets the sides. Make the necessary bend for a tight lap.
9. Complete the soldering.

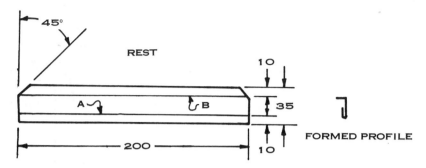

Instructions — Rest

1. Lay out and notch as shown.
2. Make a closed fold on line "A".
3. Make a square fold on line "B".
4. Solder securely with the single edge of the rest 5 mm from the back of the pan.
5. Dress all sharp edges.

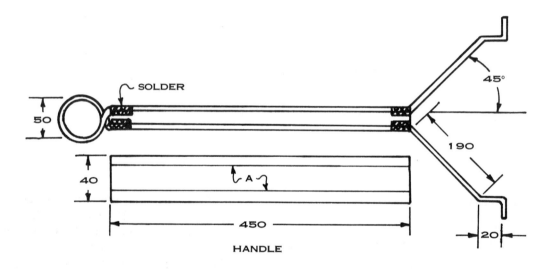

Instructions — Handle

NOTE: The wire is approximately 3 mm in diameter.

1. On a galvanized iron strip, make square folds on the lines marked "A". (The distance is determined by the wired edge allowance.)
2. Determine the middle of the wire and shape a 50 mm diameter loop.
3. Hold the crossover of wire with pliers and make sharp bends.
4. Enclose the wire with the galvanized iron strip.
5. Solder in place as shown.
6. Using pliers, make necessary bends in the wire as shown. *NOTE: Do not cut off any surplus wire.*
7. Punch holes to suit the wire in the sides of the pan, 100 mm from the back and immediately under the 5 mm fold.
8. Attach the handles. *NOTE: The pan must be permitted to swing freely within the loop.*
9. Cut off the surplus wire even with the top of the sides of the pan.

FIXED HANDLE DUST PAN

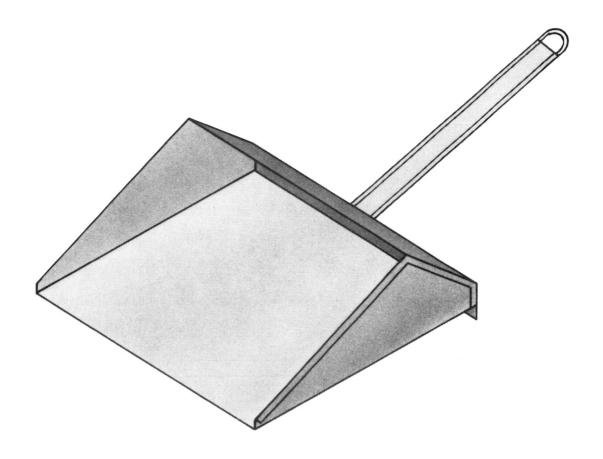

Bill of Material

Item	No. Req'd.	Material	Pre-Cut Size
Pan	1	28 Ga. galv. iron	375 mm x 225 mm
Back Rest	1	28 Ga. galv. iron	250 mm x 150 mm
Handle	1	28 Ga. galv. iron	250 mm x 50 mm
Wire	1	#10 coated	500 mm long

78 Sheet Metal Practice

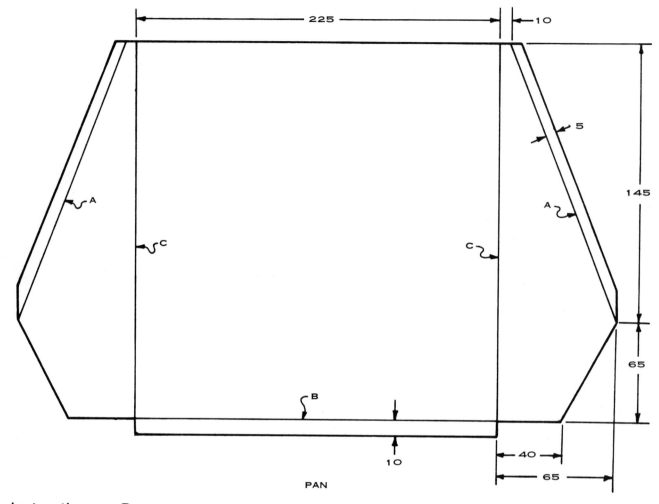

PAN

Instructions — Pan

1. Lay out as shown.
2. Cut and notch the pattern.
3. Make closed folds on the lines marked "A", away from the lines.
4. Make a square fold on line "B", in the same direction.
5. Make square bends on the lines marked "C", towards the lines.

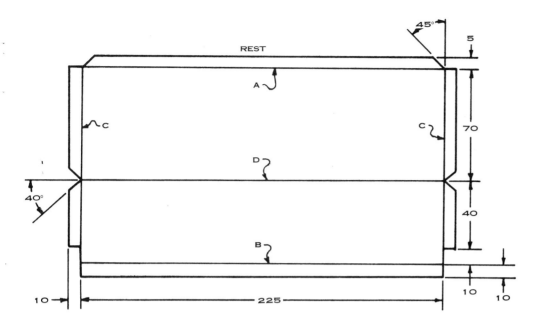

Instructions — Back Rest

1. Lay out as shown.
2. Cut and notch the pattern.
3. Make a closed fold on line "A".
   — a full fold on line "B".
   — square folds on the lines marked "C".
   — a 65° bend on line "D".

   NOTE: All folds are made towards the lines.

4. Solder the back rest neatly to the pan.

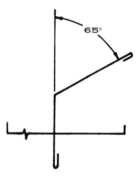

FORMED PROFILE

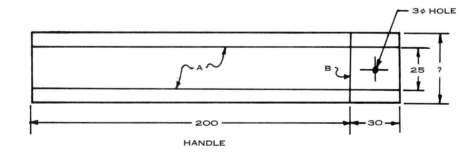

HANDLE

Instructions — Handle

1. Lay out as shown. *NOTE: The wire is approximately 3 mm in diameter.*
2. Make square folds on the lines marked "A".
3. Shape the wire to suit around a small stake.
4. Complete the wired edges.
5. Form on line "B" to 110° on a stake or in a vice.
6. Rivet and solder in place.
7. Dress all sharp edges.

CAMP CUP

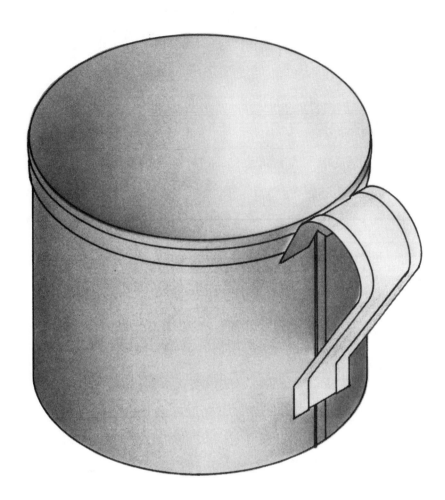

Bill of Material

Item	No. Req'd.	Material	Pre-Cut Size
Body	1	IC tin plate	250 mm x 90 mm
Bottom	1	IC tin plate	125 mm x 90 mm
Handle	1	IC tin plate	from above

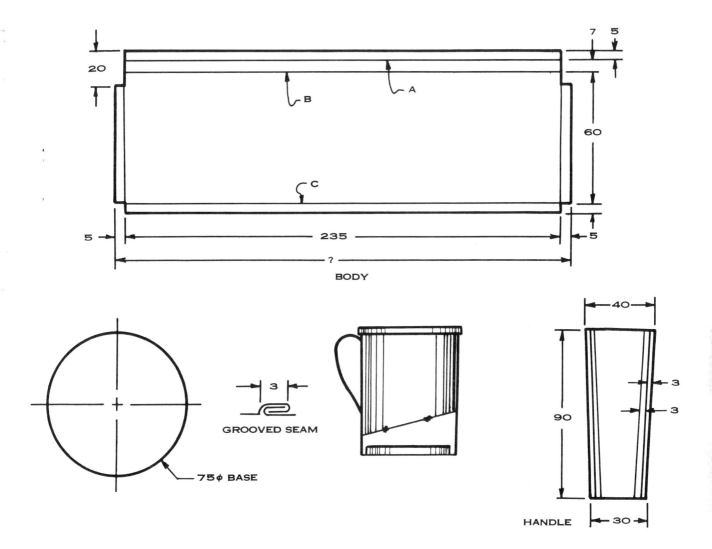

Instructions

1. Lay out and cut out patterns for the body, base, and handle.
2. Break in the body pattern.
3. Make a closed fold on line "A", towards the line.
4. Make a closed fold on line "B", in the same direction.
5. Make a closed fold on line "C", away from the line.
6. Make full folds at each end of the pattern for a 3 mm grooved seam.
7. Roll to a cylindrical shape.
8. Complete the grooved seam.
9. Make double closed folds on each side of the handle pattern.
10. Shape to suit on a suitable stake.
11. Solder the seam on the inside of the cup.
12. Solder the bottom as shown.
13. Fit the handle over the grooved seam and solder.
14. Remove all sharp edges.

HANGING PLANTER

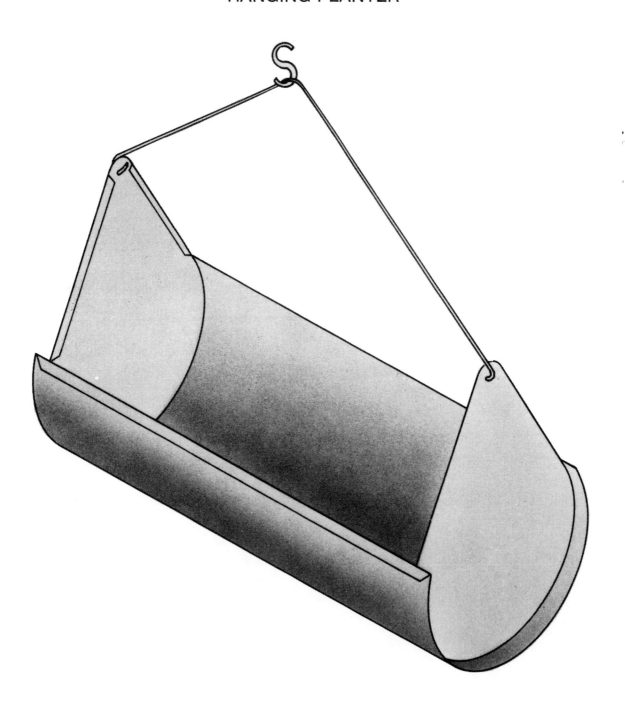

Bill of Material

Item	No. Req'd.	Material	Pre-Cut Size
Body	1	IC tin plate	250 mm x 225 mm
Ends	2	IX tin plate	175 mm x 115 mm
Hanger	1	#12 coated wire	450 mm long
"S" Hook	1	#12 coated wire	from above

Instructions — Ends

1. Lay out as shown.
2. Cut the pattern to shape.
3. Punch a 3 mm diameter hole.
4. Use the pattern as a template to produce the other end.
5. Make closed folds on the lines marked "A".

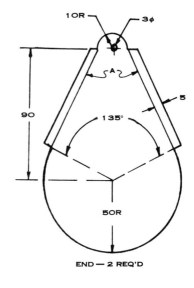

END — 2 REQ'D

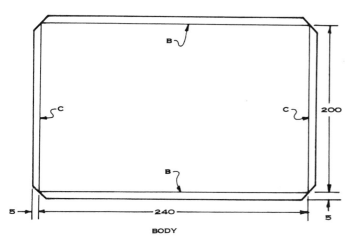

BODY

Instructions — Body

1. Lay out and notch as shown.
2. Break in the pattern.
3. Make closed folds on the lines marked "B", towards the lines.
4. Make closed folds on the lines marked "C", away from the lines.
5. Place in the roll former with the lines up and roll to the desired radius.
6. Solder the ends to the body.

Instructions — Hanger

1. Form a small "S" hook using approximately 75 mm of wire.
2. Cut off the "S" hook.
3. With the remaining material, form as shown.
4. Fit the hanger into the holes provided in the ends.

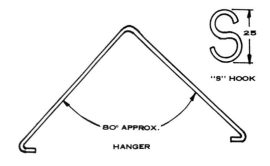

"S" HOOK

HANGER

WINDOWSILL PLANTER

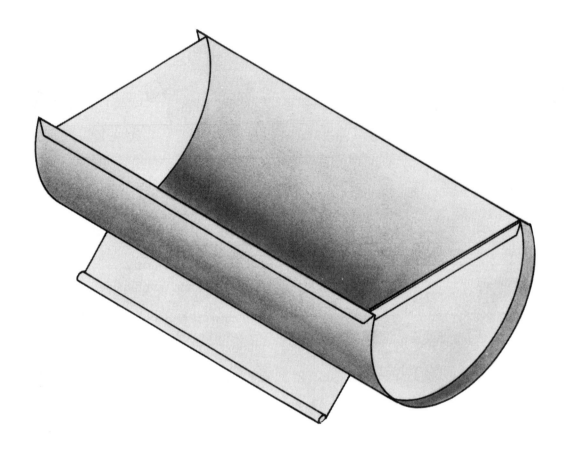

Bill of Material

Item	No. Req'd.	Material	Pre-Cut Size
Body	1	IC tin plate	250 mm x 225 mm
Ends	2	IX tin plate	125 mm x 125 mm
Supports	2	IX tin plate	175 mm x 50 mm
Wire	2	#10 coated	150 mm long

NOTE: All notches are 45°.

Instructions — Ends

1. Lay out as shown.
2. Cut the pattern to shape.
3. Use the pattern as a template to produce the other end.
4. Make a double closed fold on the lines marked "A".

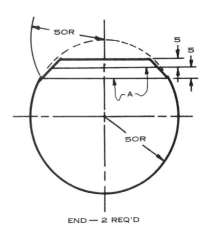

END — 2 REQ'D

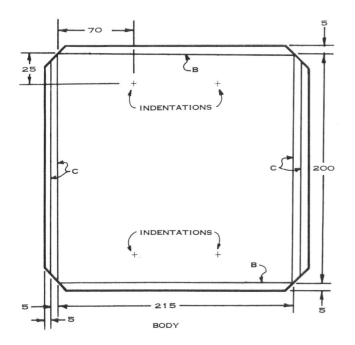

BODY

Instructions — Body

1. Lay out and notch as shown.
2. Break in the pattern.
3. Make closed folds on the lines marked "B", towards the lines.
4. Make double closed folds on the lines marked "C", away from the lines.
5. Place in the roll former with the lines up and form to the desired radius.
6. Solder the ends to the body.

Instructions — Supports

1. Lay out and notch as shown.
2. Make an open fold to 115° on line "D", towards the lines.
3. Make a square fold on line "E" away from the lines, then enclose #10 wire.
4. Locate and solder the supports.
5. Remove all sharp edges.

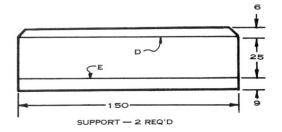

SUPPORT — 2 REQ'D

CLOTHES PEG BOX

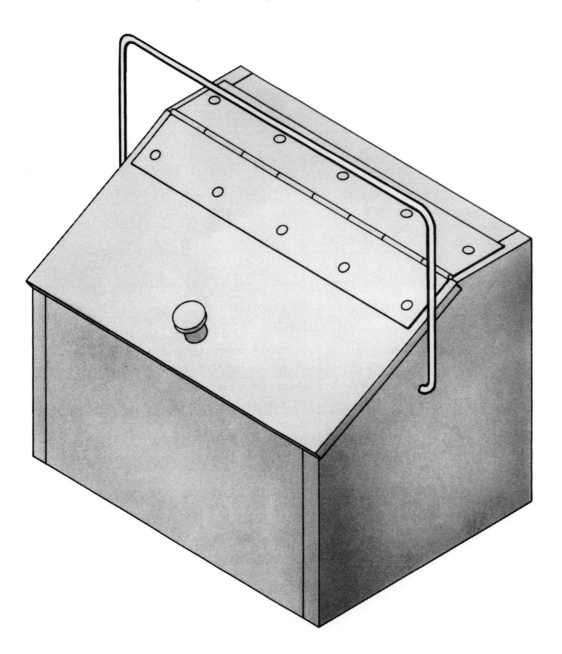

Bill of Material

Item	No. Req'd.	Material	Pre-Cut Size
Body	1	28 Ga. galv. iron	500 mm x 250 mm
Ends	2	28 Ga. galv. iron	175 mm x 175 mm
Cover	1	28 Ga. galv. iron	250 mm x 150 mm
Hinge	1	28 Ga. galv. iron	225 mm x 90 mm
Hinge Pin	1	#10 coated wire	200 mm long
Handle	1	#7 coated wire	450 mm long
Rivets	10	tinners	1 lb.

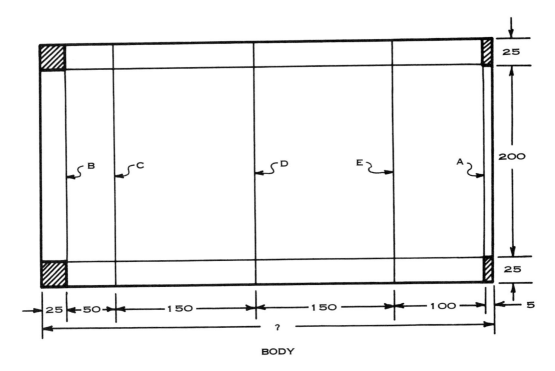

BODY

Instructions — Body

1. Lay out and notch the pattern as shown.
2. Form pockets for Pittsburgh locks.
   NOTE: If the Pittsburgh lock is formed in the standard hand brake, the first bend is made away from the line.
3. Make a closed fold on line "A", away from the line.
4. Make a closed fold on line "B", towards the line.
5. Make square bends on lines "C", "D", & "E", towards the lines.
6. Open the pockets slightly to prepare for the ends.

Instructions — Ends

1. Lay out and notch as shown.
2. Make a closed fold on line "A", away from the line.
3. Make square folds on the lines marked "B", towards the lines.
4. Insert the ends in pockets on the body.
5. Use a setting hammer to secure the ends.
6. Remove all sharp edges.

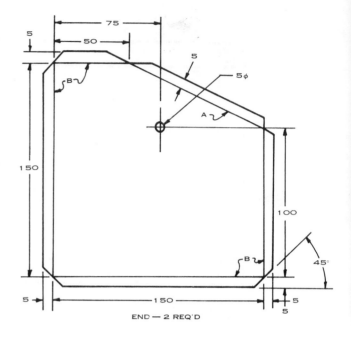

END — 2 REQ'D

Instructions — Cover

1. Lay out and notch the pattern.
2. Form 3 sides as shown in the profile.

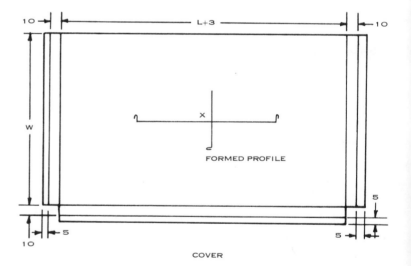

COVER

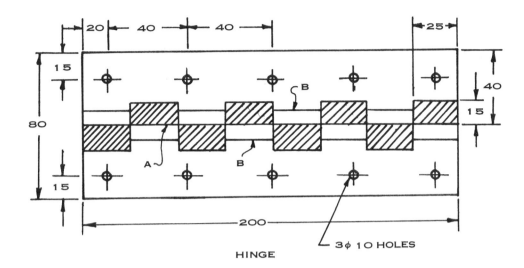

HINGE

Instructions — Hinge

1. Lay out as shown.
2. Punch 3 mm diameter holes.
3. Cut on line "A", then notch the patterns.
4. Trim the patterns so that one fits freely into the other.
5. Make square folds on the lines marked "B" to suit the wired edge allowance.

NOTE: #10 wire is approximately 3 mm in diameter.

6. Complete the hinge by enclosing the #10 wire supplied.
7. Rivet the hinge to the cover, then to the body.

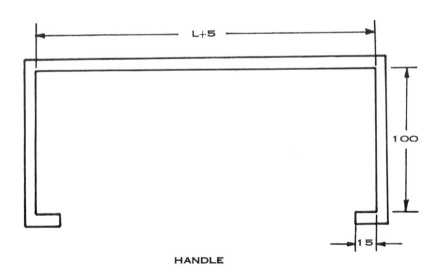

HANDLE

Instructions — Handle

1. Cut wire to the proper length, then mark the dimensions required.
2. Form in a vice or bar and tube bender, then fit in place.

CORNER SHELF

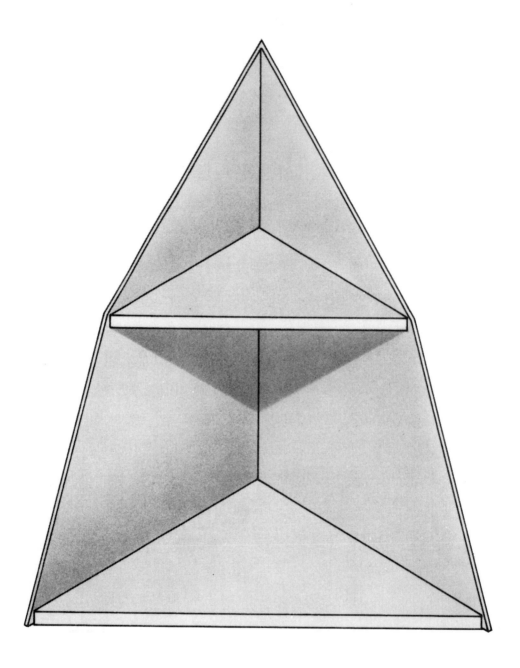

Bill of Material

Item	No. Req'd.	Material	Pre-Cut Size
Back	1	IX tin plate	500 mm x 350 mm
Top Shelf	1	IX tin plate	from above
Bottom Shelf	1	IX tin plate	from above

Instructions — Back and Shelves

1. Lay out the back and shelves as shown.
2. Cut and notch all patterns.
3. Make closed folds on all the lines marked "A", away from the line on the body, towards the line on the shelves.
4. Make a full fold on line "B", towards the line.
5. Make square folds on the lines marked "C" and "D" in that order, towards the lines.
6. Make a square bend on line "E", towards the line.
7. Place the shelves in position and solder.
8. Remove all sharp edges.

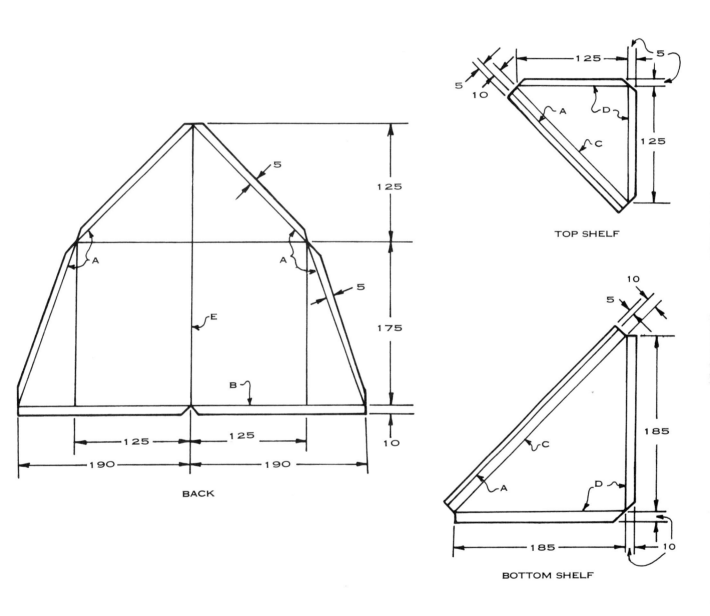

THREE RING BOOK BINDER

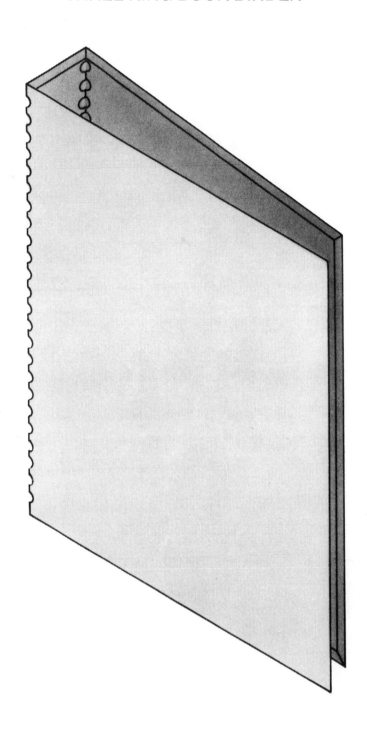

Bill of Material

Item	No. Req'd.	Material	Pre-Cut Size
Cover	2	.025 in aluminum	320 mm x 250 mm
Back	1	.025 in aluminum	320 mm x 65 mm
Hinge Pin	2	#12 coated wire	300 mm long

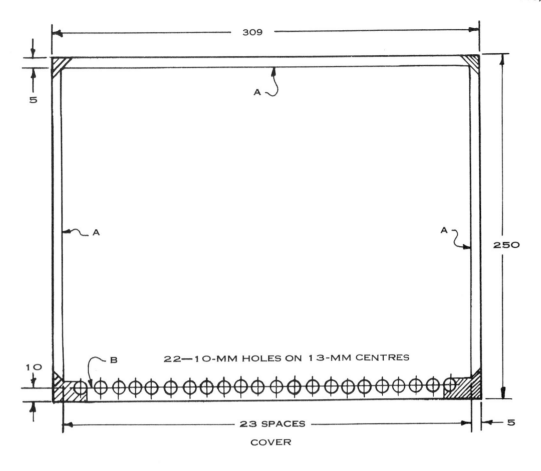

COVER

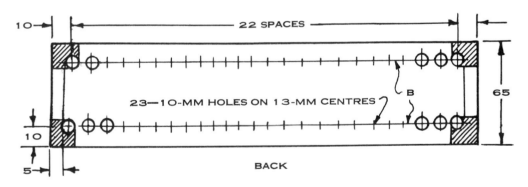

BACK

Instructions

1. Lay out as shown.
2. Centre the punch accurately for the 10 mm holes.
3. Punch the 10 mm holes.
4. Notch patterns where shown.
5. Make closed folds on lines marked "A" towards the lines.
6. Make full folds on lines marked "B", towards the lines.
7. Assemble the cover to the back and slide in the #12 wire.
8. Close the full folds on the hinge using a #3 hand groover.
9. Remove the ring assembly from an old three-ring binder, being careful to salvage the rivets.
10. Align the rings on the back, then drill and rivet in place.
11. Dress all sharp edges.

COAL SCUTTLE ASH TRAY

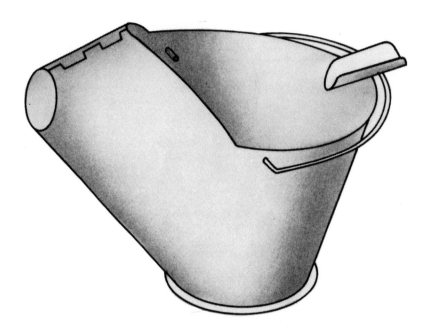

Bill of Material

Item	No. Req'd.	Material	Pre-Cut Size
Body	1	light gauge copper	250 mm x 150 mm
Bottom	1	light gauge copper	from above
Rest	1	light gauge copper	from above
Handle	1	14 Ga. copper wire	250 mm long

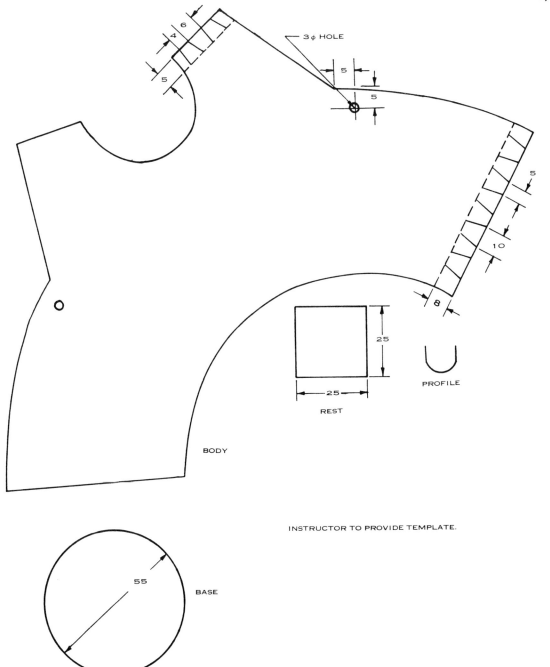

INSTRUCTOR TO PROVIDE TEMPLATE.

Instructions — Body, Base, Handle, and Rest

1. Trace the body pattern on the metal and cut to shape.
2. Lay out dovetail seams and holes for the handle. *NOTE: When cutting dovetails, no material is removed.*
3. Lay out and cut patterns for the base and rest.
4. Form the body to shape by hand or over a suitable stake.
5. On the long seam, bend alternate dovetails up to 90°, slip the plain side in, tap down, then solder the seam.
6. Repeat step 5 for the short seam.
7. Form the handle to suit and solder the rest in place.
8. Attach the handle so that it swings freely.
9. Polish all surfaces and spray with clear lacquer.

AQUARIUM

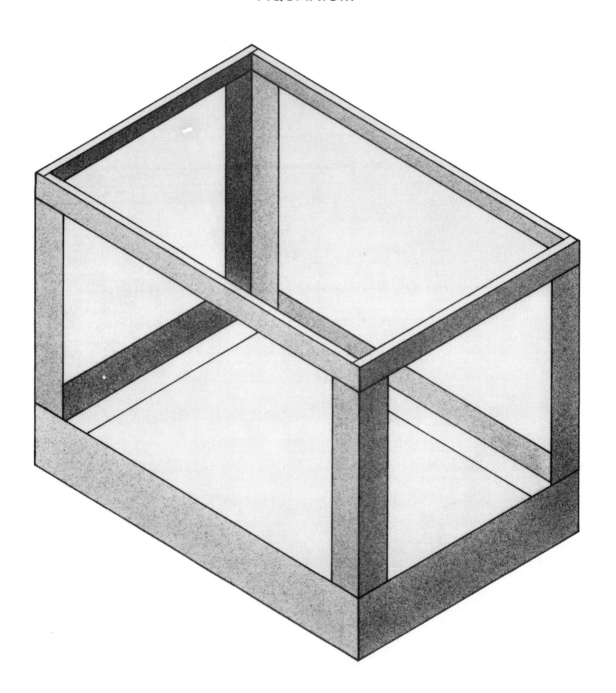

Bill of Material

Item	No. Req'd.	Material	Pre-Cut Size
Side Bottom Rails	2	26 Ga. galv. iron	750 mm x 300 mm
End Bottom Rails	2	26 Ga. galv. iron	from above
Side Top Rails	2	26 Ga. galv. iron	from above
End Top Rails	2	26 Ga. galv. iron	from above
Vertical Angles	4	26 Ga. galv. iron	from above

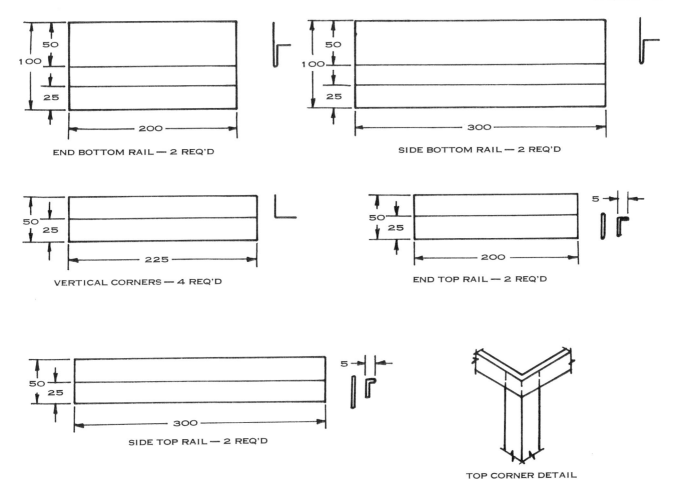

Instructions

1. Cut all pieces to the correct sizes shown.
2. Lay out each piece individually.
3. Form as shown in the formed profiles.
4. Fit the pieces together as shown in details.

   NOTE: Square all corners before soldering.

5. Solder all joints securely.

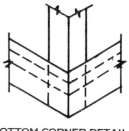

FISHING TACKLE BOX

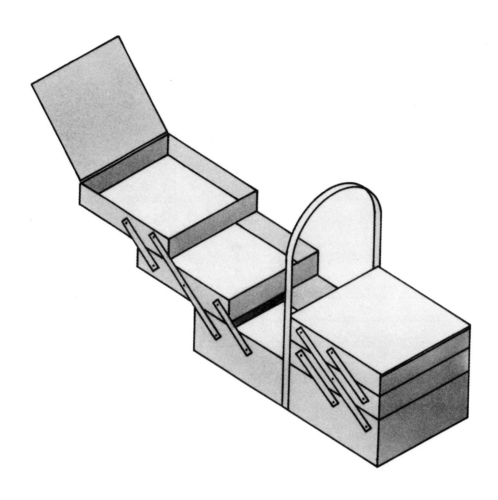

Bill of Material

Item	No. Req'd.	Material	Pre-Cut Size
Top Trays	2	IX tin plate	250 mm x 250 mm
Middle Trays	2	IX tin plate	250 mm x 250 mm
Bottom Tray	1	IX tin plate	450 mm x 300 mm
Covers	2	IX tin plate	175 mm x 175 mm
Hinge Pins	2	#12 wire	150 mm long
Hinge Bars	4	2 mm x 10 mm	150 mm long
Hinge Bars	8	2 mm x 10 mm	80 mm long
Handle	1	2 mm x 20 mm	600 mm long
Rivets	28	3 mm diam. split	10 mm long

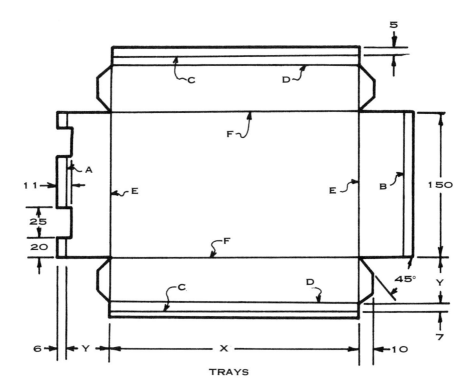

TRAYS

NOTE: Make hinge notches in the top trays only.

Instructions — Top Trays
X = 150 mm, Y = 30 mm

1. Lay out and notch the pattern. *NOTE: All folds are made towards the lines.*
2. Make a square fold on line "A".
3. Enclose the wire in the fold.
4. Make closed folds on the lines marked "B" and "C".
5. Make square folds on the lines marked "D".
6. Complete forming by making square bends on the lines marked "E" and "F".

NOTE: Laps are to be placed inside.

7. Solder or rivet seams as required.

Instructions — Middle Trays
X = 150 mm, Y = 30 mm

1. Lay out and notch the pattern.
2. Make a closed fold on line "A".
3. Repeat steps 4, 5, 6, and 7 above.

Instructions — Bottom Tray
X = 305 mm, Y = 65 mm

1. Complete steps 1, 2, and 3 of the middle trays.

Instructions — Covers

1. Lay out and notch the patterns.
2. Make closed folds on the lines marked "A".
3. Make a square fold on line "B".

NOTE: Place the tabs for the hinge under the wire on the top tray and bend around the wire after the hinge bars are riveted in place.

Instructions — Handle

1. Rivet or bolt in place.

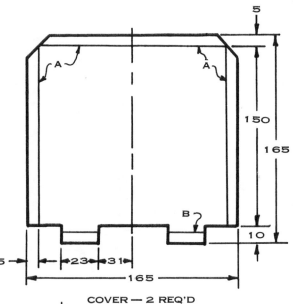

COVER — 2 REQ'D

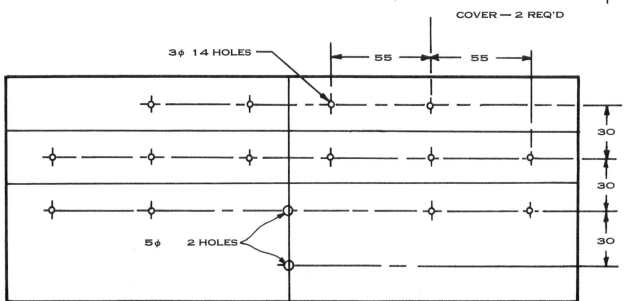

DRILLING LAYOUT FOR HINGE BAR AND HANDLE MOUNTING HOLES

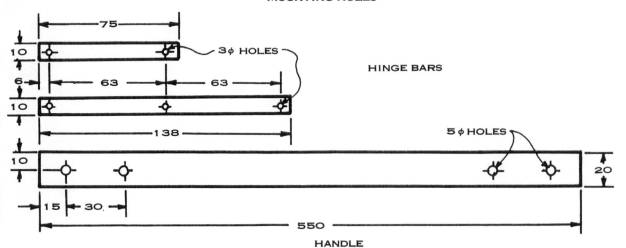

HINGE BARS

HANDLE

WATERING CAN

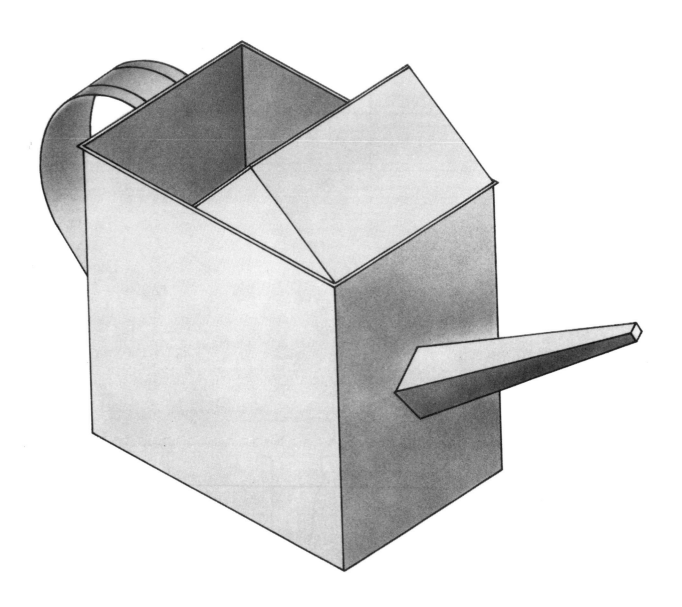

Bill of Material

Item	No. Req'd.	Material	Pre-Cut Size
Body	1	28 Ga. galv. iron	525 mm x 175 mm
Bottom	1	28 Ga. galv. iron	175 mm x 125 mm
Hood	1	28 Ga. galv. iron	175 mm x 100 mm
Handle	1	28 Ga. galv. iron	225 mm x 50 mm
Spout	1	28 Ga. galv. iron	225 mm x 125 mm
Wire	1	#12 coated	525 mm long
Wire	1	#12 coated	200 mm long
Wire	2	#10 coated	225 mm long

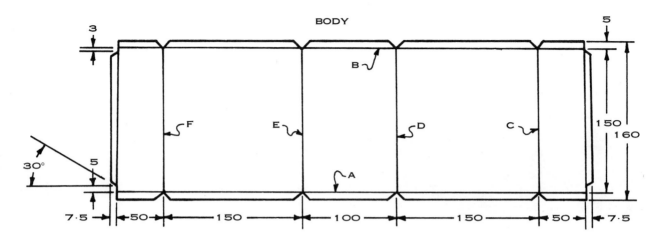

Instructions — Body

1. Lay out and notch as shown.
2. Make proper folds on each end for a 5 mm grooved seam.
3. Make a full fold on line "A", towards the line.
4. Make a square fold on line "B", away from the line.
5. Make square folds on lines "C", "D", "E", and "F", in order.
6. Complete the grooved seam.
7. Complete the wired edge.

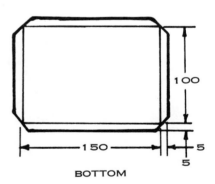

Instructions — Bottom

1. Lay out and notch as shown.
2. Make 5 mm square folds on all sides.
3. Solder in place.

Instructions — Hood

1. Lay out and notch as shown.
2. Make a square fold on line "A", away from the line.
3. Make square bends on the lines marked "B", towards the lines.
4. Complete the wired edge.
5. Solder in place after the spout has been located.

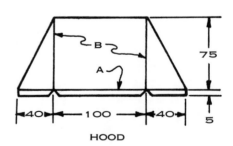

Instructions — Handle

1. Lay out and make square folds on the lines marked "A" to suit #10 wire.

   NOTE: #10 wire is approximately 3 mm.

2. Enclose the wire in the folds.
3. Form to the desired shape in the roll former, then solder in place.

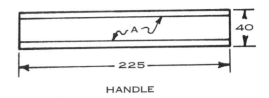

HANDLE

Instructions — Spout

1. Lay out and cut as shown.
2. Form to the desired shape where necessary.
3. Make an opening in the body to suit, and solder in place.

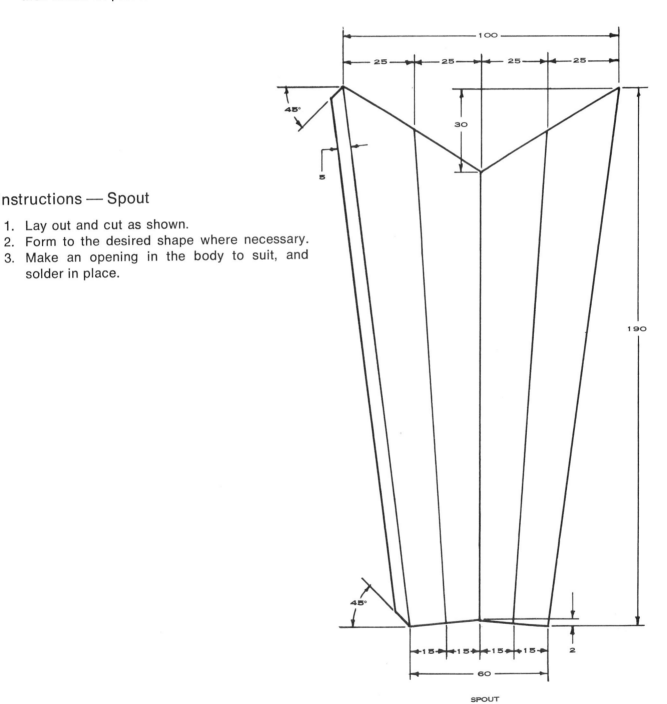

SPOUT

SHOESHINE BOX

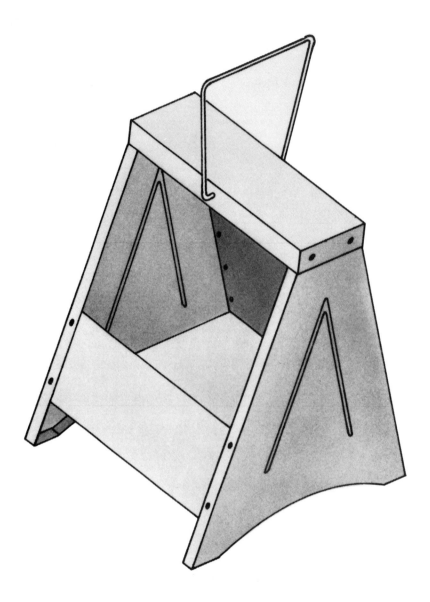

Bill of Material

Item	No. Req'd.	Material	Pre-Cut Size
Body	1	26 Ga. galv. iron	600 mm x 260 mm
Ends	2	26 Ga. galv. iron	325 mm x 300 mm
Top	1	light gauge aluminum	300 mm x 150 mm
Handle	1	#6 coated wire	450 mm long
Rivets	12	1 lb. tinners	
Sheet Metal Screws	4	#8 binding head x 10 mm long	

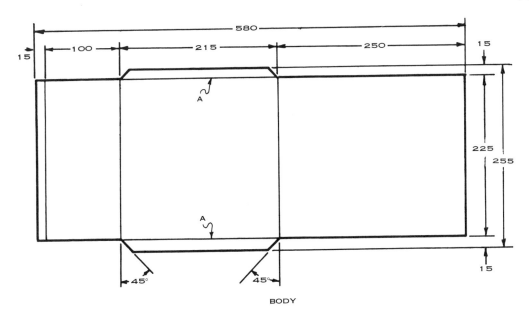

BODY

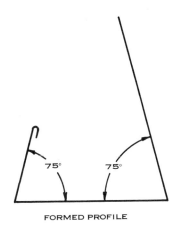

FORMED PROFILE

Instructions — Body

1. Lay out and notch as shown.
2. Make square folds on the lines marked "A", away from the lines.
3. Form as shown in the profile with all folds towards the lines.

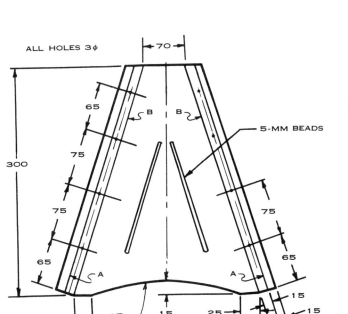

END — 2 REQ'D

TURN 5MM EDGE INSIDE ON ARC

ALL HOLES 3⌀

Instructions — Ends

1. Lay out and cut the pattern as shown.
2. Punch or drill all 3 mm diameter holes.
3. Use the pattern as a template to produce the other end.
4. Make 5 mm beads where shown, away from the lines.
5. Turn a 5 mm edge on the arc, towards the lines.
6. Make closed folds on the lines marked "A", towards the lines.
7. Make square folds on the lines marked "B", towards the lines.
8. Locate and punch holes in the body, then rivet in place.

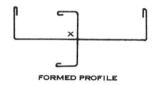
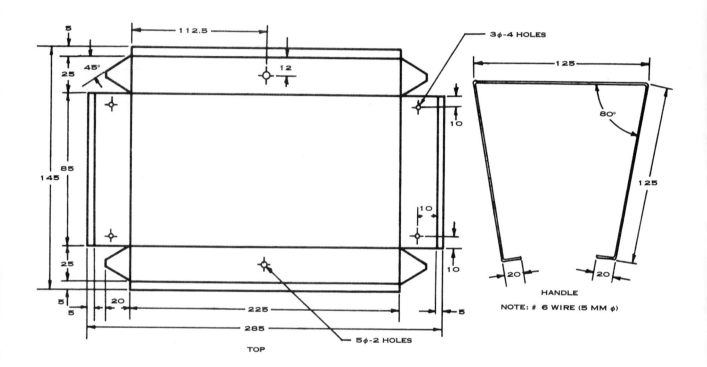

Instructions — Top

1. Lay out and notch the pattern as shown.
2. Punch or drill all holes.
3. Form as shown in the profile with the laps inside.
4. Place in position and drill through the ends.
5. Secure the top with sheet metal screws.

Instructions — Handle

1. Lay out as shown using the inside dimensions.
2. Form to shape in a vice or wire bender.
3. Insert in the 5 mm diameter holes in the top.

FILING CABINET

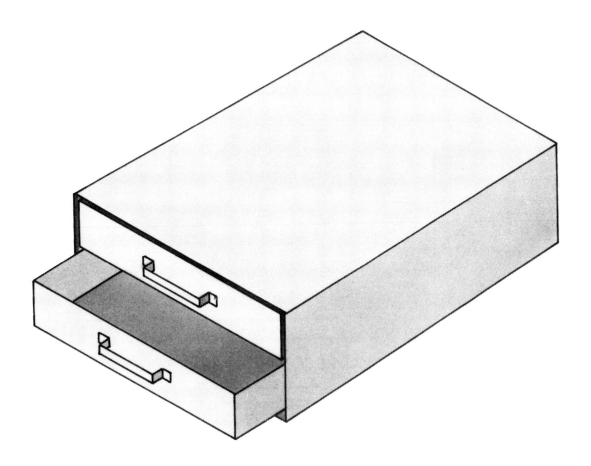

Bill of Material

Item	No. Req'd.	Material	Pre-Cut Size
Body	1	IX tin plate	500 mm x 450 mm
Bottom	1	IX tin plate	350 mm x 250 mm
Drawers	2	IX tin plate	400 mm x 350 mm
Runners	2	IX tin plate	300 mm x 75 mm
Handles	2	own design	

108 Sheet Metal Practice

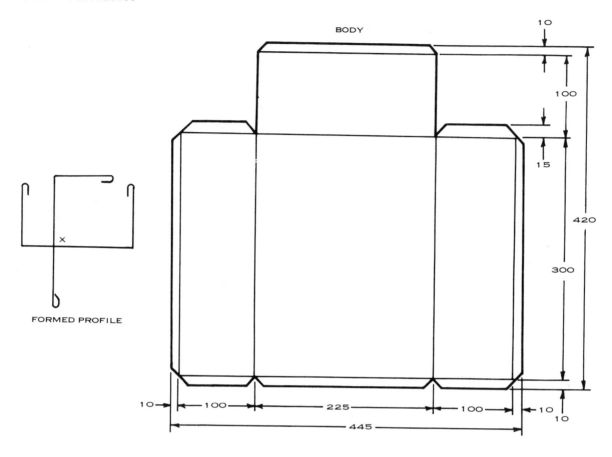

Instructions — Body

1. Lay out and notch the pattern as shown.

   NOTE: All angle notches are 45°.

2. Form as shown in the profile, with all folds towards the inside.
3. Solder laps on the inside.

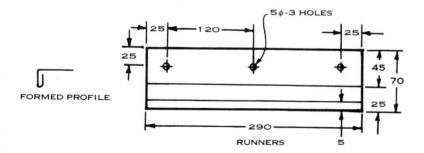

Instructions — Runners

1. Lay out and punch the holes as shown.
2. Form as shown in the profile.
3. Locate and solder through 5 mm holes.

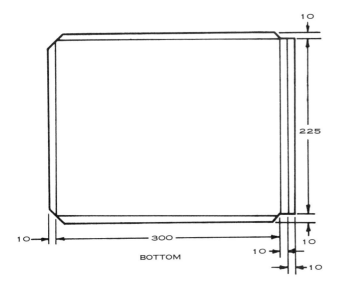

BOTTOM

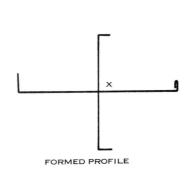

FORMED PROFILE

Instructions — Bottom

1. Lay out and notch as shown.
2. Form as shown in the profile.
3. Position in the body and tack solder in place.

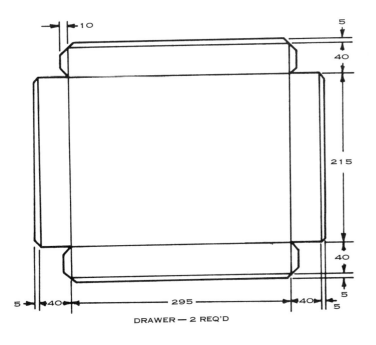

DRAWER — 2 REQ'D

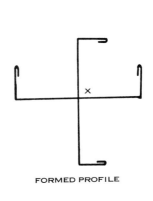

FORMED PROFILE

Instructions — Drawers

1. Lay out and cut the patterns as shown.
2. Form as shown in the profile with the laps on the inside.
3. Solder the corners neatly.
4. Make handles of your own design and attach them to the drawers.

110 Sheet Metal Practice

FILING TRAY

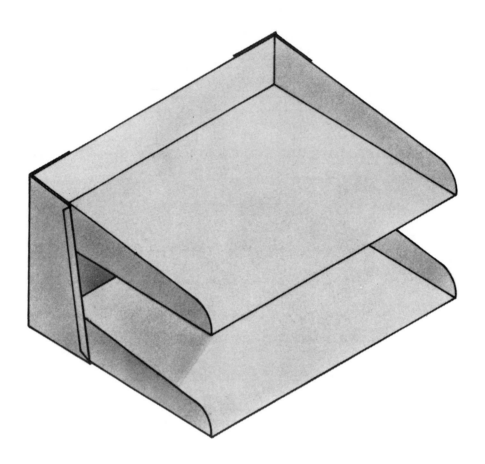

Bill of Material

Item	No. Req'd.	Material	Pre-Cut Size
Tray	2	24 Ga. Satin Coat	425 mm x 300 mm
Support	2	24 Ga. Satin Coat	175 mm x 150 mm

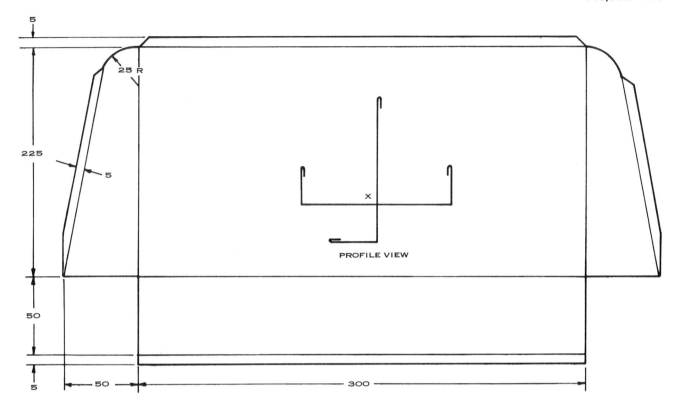

Instructions — Trays

NOTE: All notches are 45° unless otherwise indicated.

1. Lay out and notch as shown.
2. Form as per profile view.

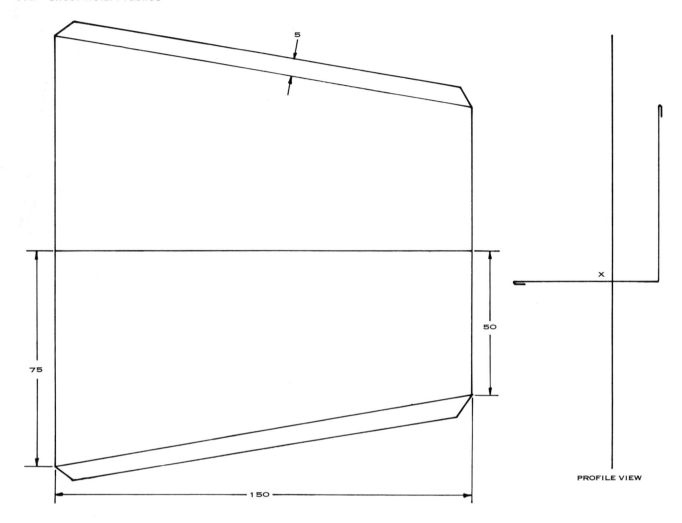

Instructions — Supports

NOTE: All notches are 45° unless otherwise indicated.

1. Lay out and notch as shown.
2. Form as per profile view.
3. Spotweld or rivet trays to supports.
4. Dress all sharp edges.

CARRY ALL

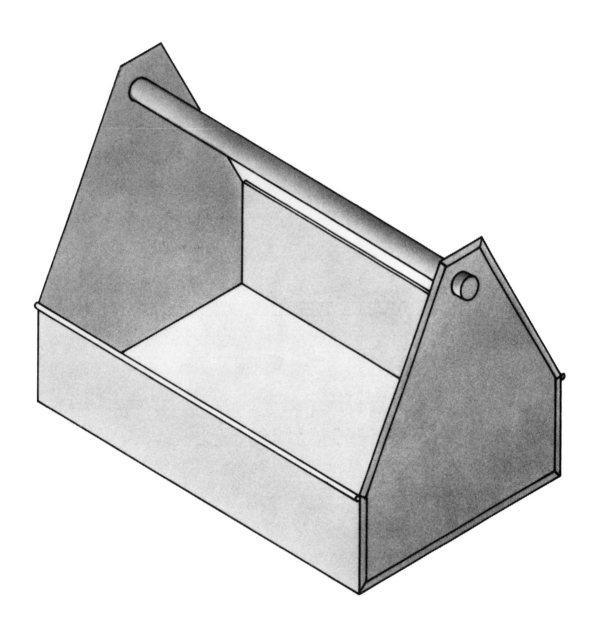

Bill of Material

Item	No. Req'd.	Material	Pre-Cut Size
Body	1	26 Ga. galv. iron	400 mm x 375 mm
Ends	2	26 Ga. galv. iron	225 mm x 225 mm
Wire	2	#6 coated	750 mm long
Handle	1	½ in. E.M.T. tubing	350 mm long
	2	½ in. E.M.T. set screw connectors	

114   Sheet Metal Practice

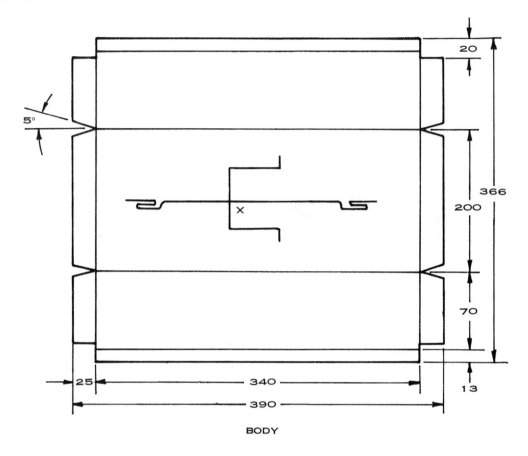

BODY

Instructions — Body

1. Lay out and notch as shown.
2. Form the pockets for the Pittsburgh locks.
3. Complete forming per profile view.

Instructions — Ends

1. Lay out and notch as shown.
2. Punch the 22 mm diameter hole.
3. Form as per profile view.

Instructions — Assembly

1. Insert ends in pockets and hammer down flange.
2. Form, fit, cut, and secure wire in place. Cut and attach handle and connectors.
3. Dress all sharp edges.

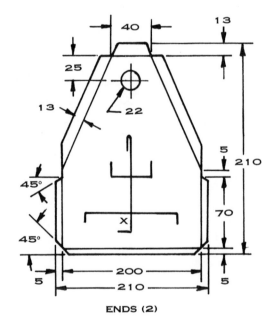

ENDS (2)

TOOL BOX

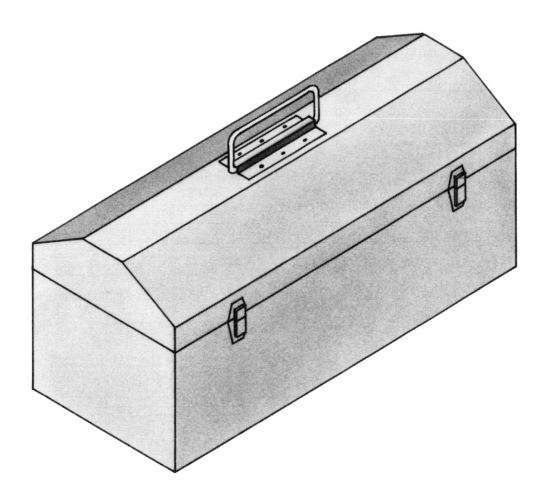

Bill of Material

Item	No. Req'd.	Material	Pre-Cut Size
Body	1	26 Ga. galv. iron	500 mm x 450 mm
Ends	2	26 Ga. galv. iron	200 mm x 175 mm
Cover	1	26 Ga. galv. iron	450 mm x 300 mm
Ends	2	26 Ga. galv. iron	200 mm x 90 mm
Hinge	1	plated steel	375 mm x 20 mm
Trunk Catches	2	plated steel	
Handle	1	own design	
Rivets (pull-stem)	as req'd.	hollow steel	⅛ in. dia. x 5/16 in.

116 Sheet Metal Practice

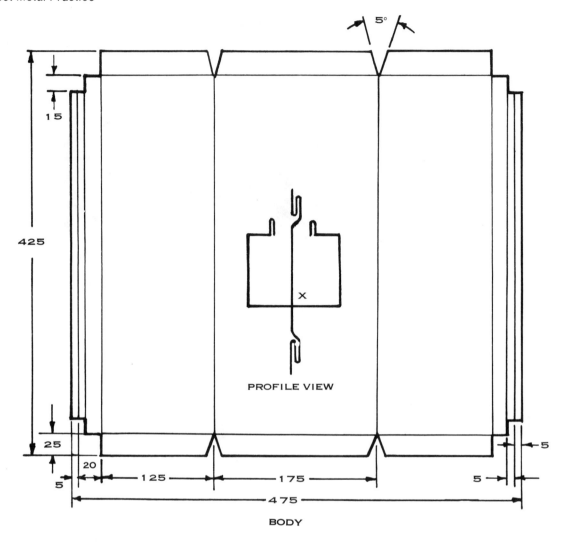

BODY

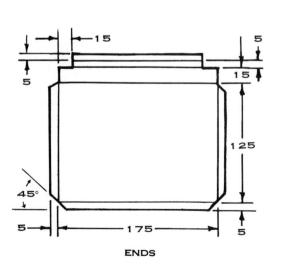

ENDS

Instructions — Body and Ends

1. Lay out and notch as shown.
2. Form as per profile views.
3. Insert ends in pockets and hammer down flange.

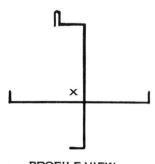

PROFILE VIEW

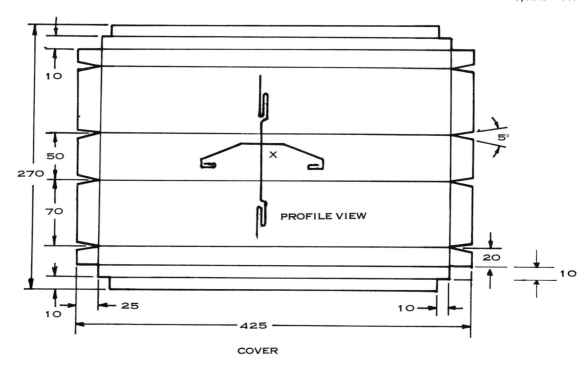

COVER

Instructions — Cover and Ends

1. Lay out and notch as shown.
2. Form as per profile views.
3. Insert ends in pockets and hammer down flange.
4. Locate, drill and rivet hinge in place.
5. Locate, drill and rivet trunk catches in place.
6. Locate, drill and rivet handle in place.
7. Dress all sharp edges.

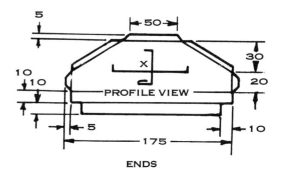

ENDS

BARBECUE

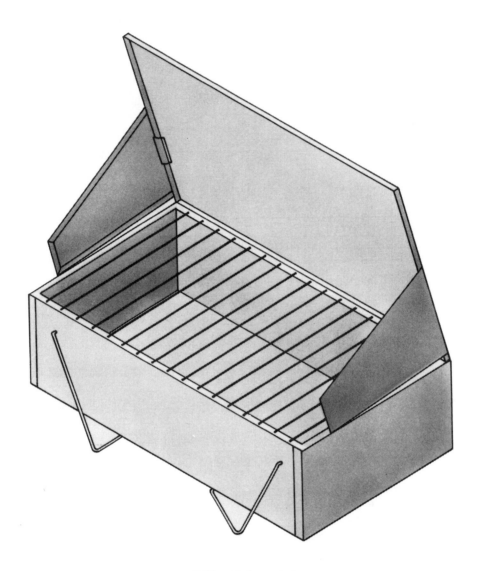

Bill of Material

Item	No. Req'd.	Material	Pre-Cut Size
Casing	1	26 Ga. galv. iron	550 mm x 450 mm
Casing ends	2	26 Ga. galv. iron	275 mm x 150 mm
Liner	1	26 Ga. galv. iron	600 mm x 450 mm
Fire box	1	28 Ga. black iron	425 mm x 275 mm
Fire box brackets	2	28 Ga. black iron	200 mm x 50 mm
Cover	1	.025 utility aluminum	425 mm x 300 mm
Baffles	2	.025 utility aluminum	225 mm x 150 mm
Hinge	2	28 Ga. galv. iron	100 mm x 35 mm
Hinge pin	1	#10 galv. wire	400 mm long
Baffle hinge pin	2	#10 galv. wire	250 mm long
Comb. legs and handles	2	#6 galv. wire	800 mm long
Grille frame	1	#6 galv. wire	1250 mm long
Grille cross members	20	#10 galv. wire	250 mm long

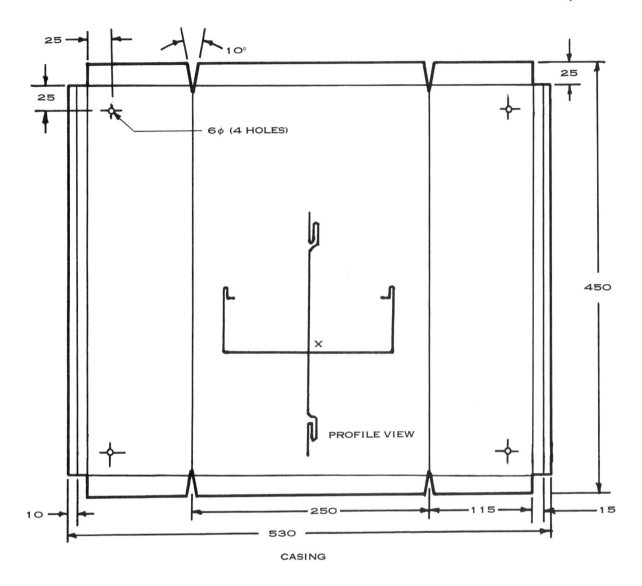

CASING

Instructions — Casing

1. Lay out and notch as shown.
2. Punch the 6 mm diameter holes.
3. Form the pockets for the Pittsburgh locks.
4. Complete forming as per profile view.

120 Sheet Metal Practice

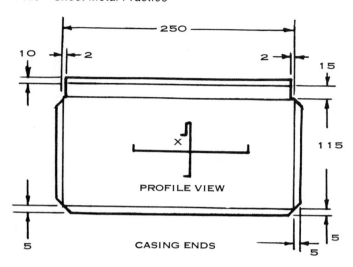

Instructions — Casing Ends

1. Lay out and notch as shown.
2. Form as per profile view.
3. Insert ends in pockets and hammer down flange.

Instructions — Liner

1. Lay out and notch as shown.
2. Punch the 3 mm diameter holes.
3. Complete forming as per profile view.
4. Drill and rivet to casing.

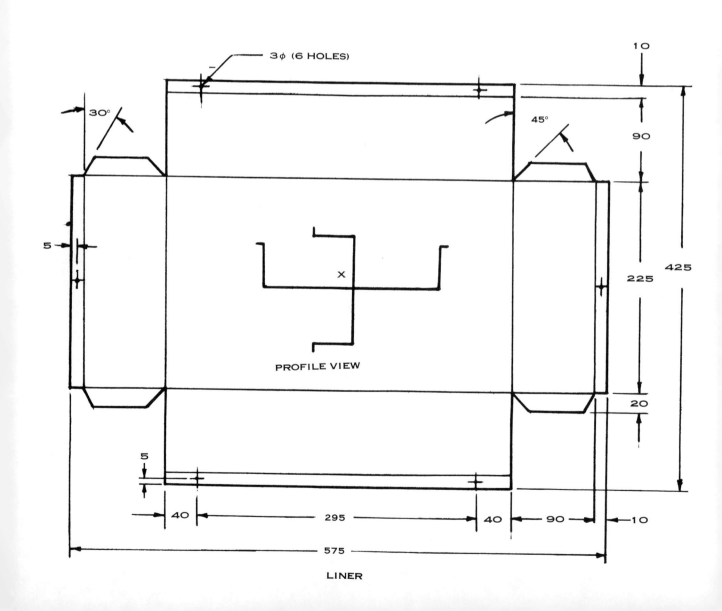

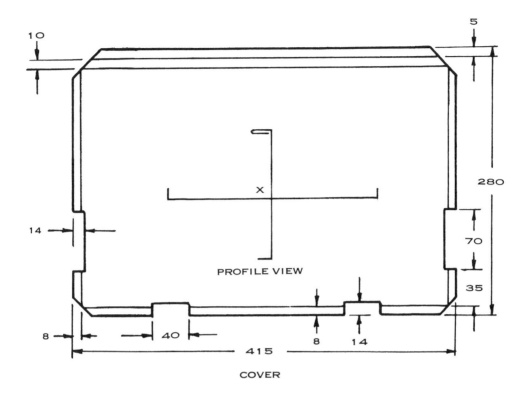

COVER

Instructions — Cover

1. Lay out and notch as shown.
2. Form as per profile view.
3. Insert hinge pins and complete wired edges.

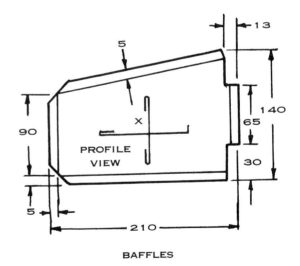

BAFFLES

Instructions — Baffles

1. Lay out and notch as shown.
2. Form as per profile view.
3. Attach to cover.

122 Sheet Metal Practice

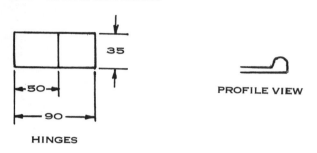

Instructions — Hinges

1. Lay out and form as shown.
2. Slip over hinge pin.
3. Punch two 3 mm diameter holes in each for mounting.
4. Locate cover and drill matching 3 mm holes in casing.
5. Pop rivet hinges to casing.

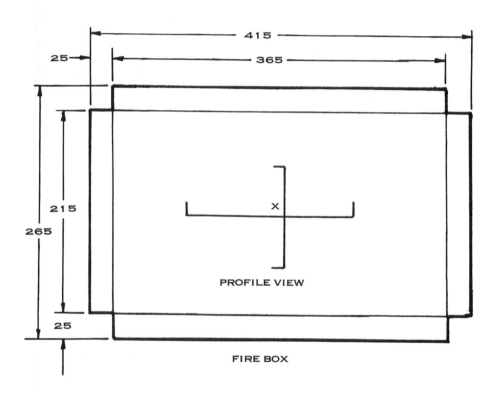

Instructions — Fire Box

1. Lay out and notch as shown.
2. Form as per profile view.

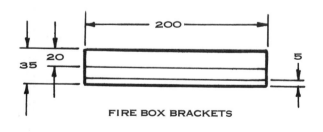

Instructions — Fire Box Brackets

1. Lay out as shown.
2. Form as per profile view.
3. Spot weld on fire box bottom, 5 mm in from edge.

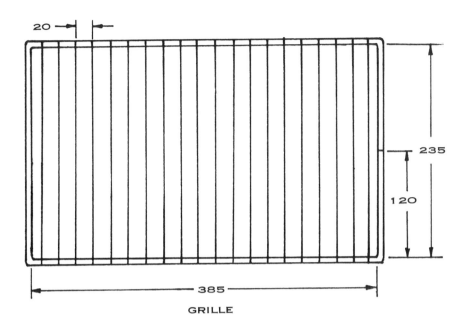

Instructions — Grille

1. Lay out frame as shown.
2. Shape in the bar and tube bender or in a vise.
3. Spot weld cross members at 20 mm centres.
4. Trim and file cross members.

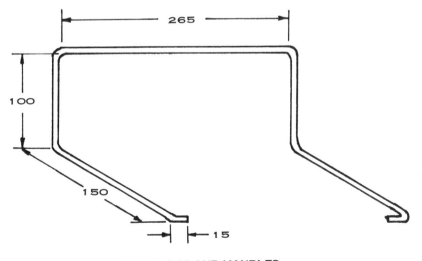

COMB. LEGS AND HANDLES

Instructions — Combination Legs and Handles

1. Lay out as shown.
2. Shape in the bar and tube bender or a vise.
3. Insert ends in 6 mm diameter holes in casing.

PICNIC COOLER

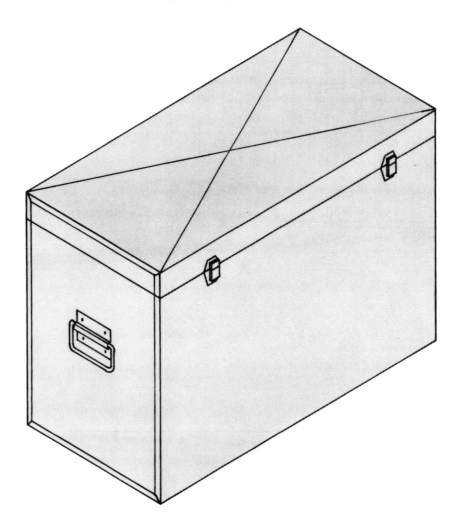

Bill of Material

Item	No. Req'd.	Material	Pre-Cut Size
Casing	1	26 Ga. galv. iron	900 mm x 500 mm
Casing ends	2	26 Ga. galv. iron	325 mm x 300 mm
Liner	1	26 Ga. galv. iron	750 mm x 450 mm
Liner ends	2	26 Ga. galv. iron	275 mm x 250 mm
Cover	1	26 Ga. galv. iron	500 mm x 450 mm
Cover ends	2	26 Ga. galv. iron	285 mm x 85 mm
Cover liner	1	26 Ga. galv. iron	500 mm x 310 mm
Handles	2	Own choice	
Piano hinge	1	Plated steel	450 mm x 40 mm
Trunk catches	2	Plated steel	
Insulation	2	25 mm fiber glass	450 mm x 275 mm
Insulation	2	25 mm fiber glass	450 mm x 275 mm
Insulation	2	25 mm fiber glass	250 mm x 225 mm
Rivets, pull-stem	20 (approx.)	Steel	⅛ in. dia. x ⁵⁄₁₆ in.

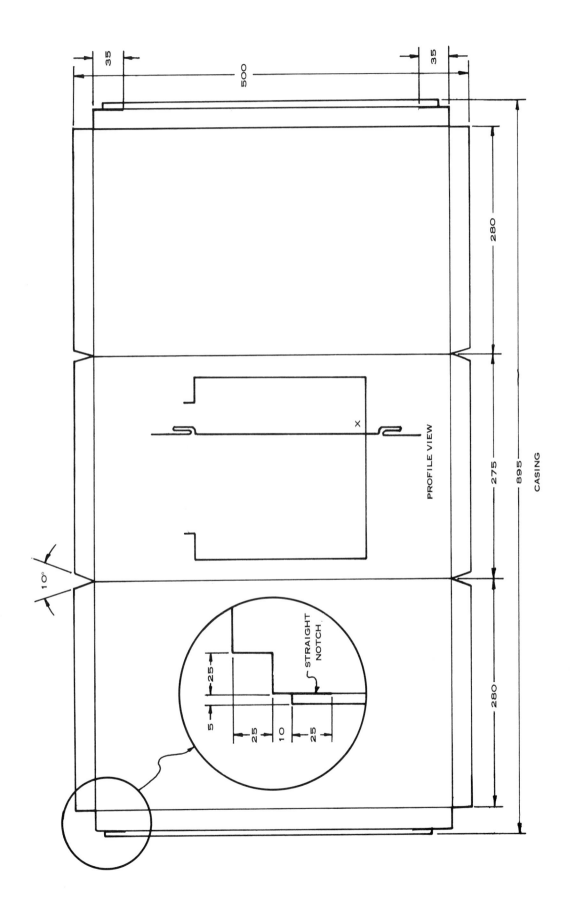

Instructions — Casing

1. Lay out and notch as shown.
2. Form pockets for Pittsburgh locks.
3. Complete the forming as per profile view.

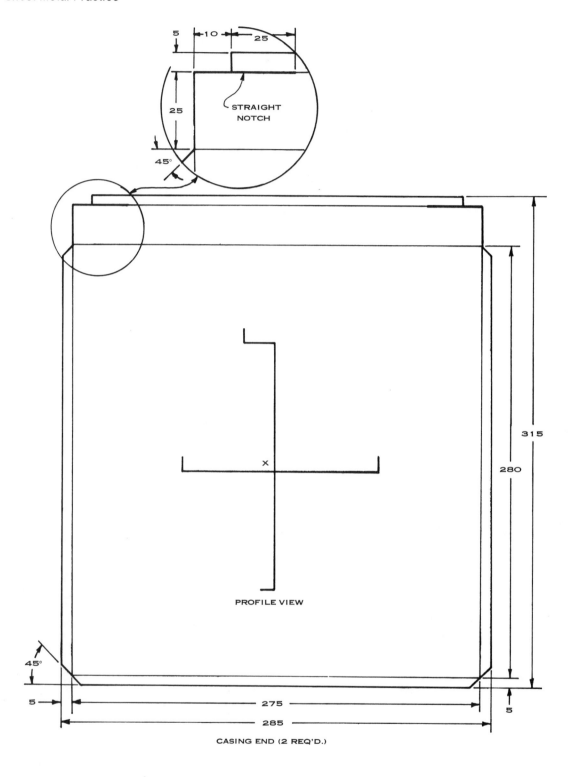

CASING END (2 REQ'D.)

Instructions — Casing End

1. Lay out and notch as shown.
2. Complete the forming as per profile view.
3. Rivet handles of own design near top in the centre of each end (a back plate may be necessary).
4. Insert ends in pockets and hammer down flange.
5. Solder Pittsburgh lock seams.

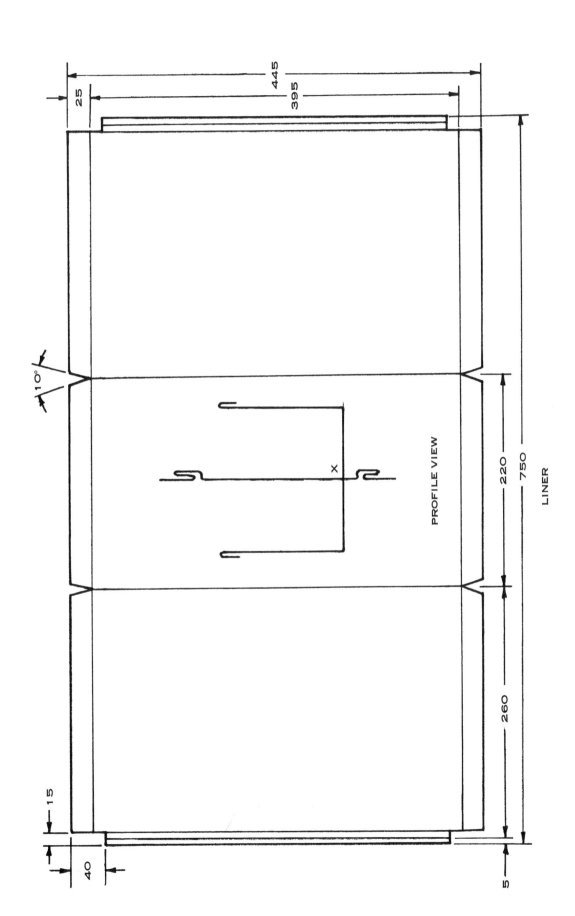

Instructions — Liner

1. Lay out and notch as shown.
2. Form pockets for Pittsburgh locks.
3. Complete the forming as per profile view.

128 Sheet Metal Practice

LINER END (2 REQ'D.)

Dimensions shown: 20, 15, 5, 260, 270, 220, 230, 5, 45°

PROFILE VIEW

Instructions — Liner End

1. Lay out and notch as shown.
2. Form as per profile view.
3. Insert ends in pockets and hammer down flange.
4. Solder Pittsburgh lock seams.

Assembly of Casing and Liner

1. Fit fiber glass inside casing to cover all surfaces.
2. Insert the liner in place.
3. Squeeze the top edge together and solder all around.
4. With the extra material at each corner, shape to a 15 mm radius (approx.) then solder and fill in where necessary.

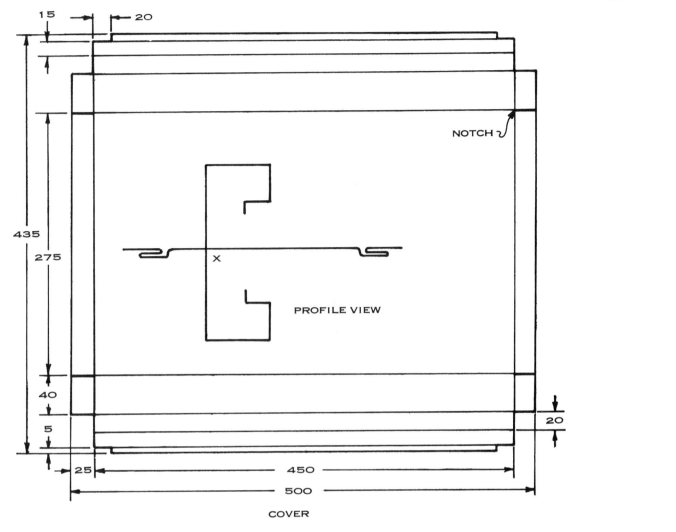

COVER

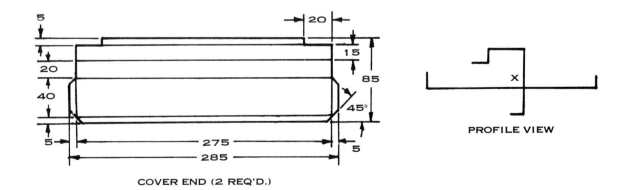

COVER END (2 REQ'D.)

Instructions — Cover

1. Lay out and notch as shown.
2. Form pockets for Pittsburgh locks.
3. Complete the forming as per profile view.

Instructions — Cover Ends

1. Lay out and notch as shown.
2. Form as per profile view.

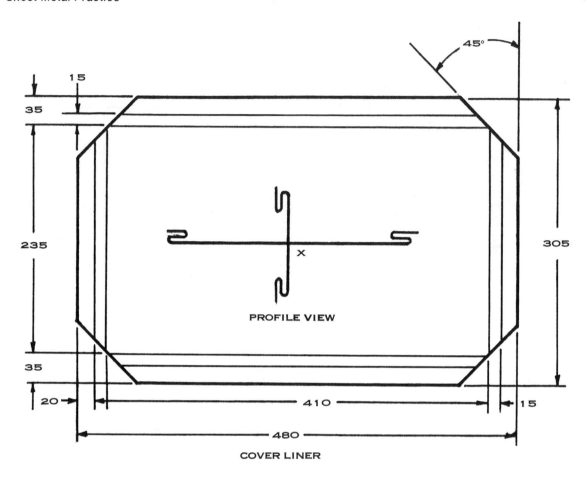

Instructions — Cover Liner

1. Lay out and notch as shown.
2. Form as per profile view.

Instructions — Cover Assembly

1. Insert one end in pocket and hammer down flange.
2. Place fiber glass in cover.
3. Slide cover liner in place.
4. Secure other end and solder.

Instructions — Final Assembly

1. Drill and "pop" rivet piano hinge to cover.
2. Line up, drill and "pop" rivet piano hinge to casing.
3. Position, drill and "pop" rivet trunk catches.
4. File and dress all sharp edges.

NOTE: To seal it may be necessary to install a "U" - shaped neoprene gasket on joining edge of casing and liner.

FUNNEL

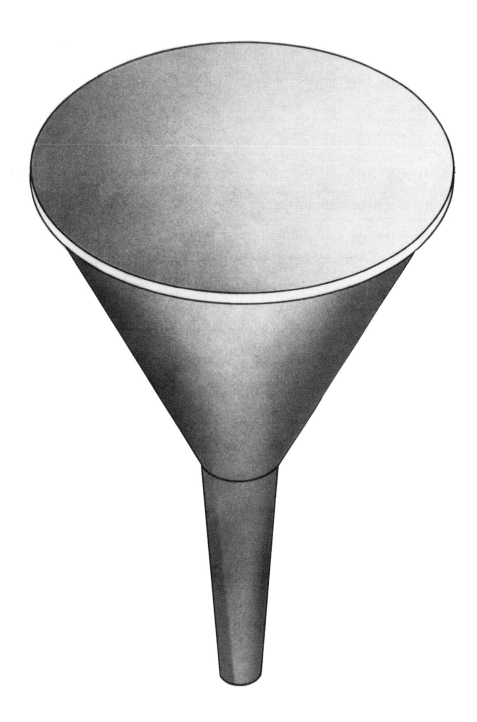

Bill of Material

Item	No. Req'd.	Material	Pre-Cut Size
Body	1	1C tin plate	350 mm x 250 mm
Spout	1	1C tin plate	from above
Wire	1	#10 coated	400 mm long

132 Sheet Metal Practice

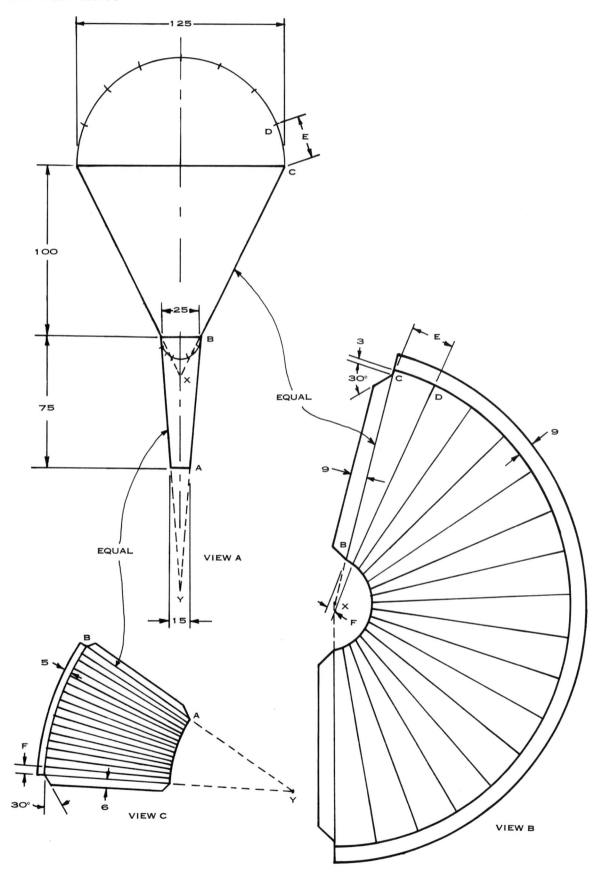

Instructions — Body

1. Draw View A, Elevation and Half Plan, full size, on paper.
2. Lay out and cut View B.
3. Break in.
4. Make full folds for 6 mm grooved seam.
5. Shape the body on a suitable stake.
6. Complete the grooved seam.
7. Turn 9 mm edge out about 90°.
8. Roll wire to shape and complete wired edge.

Instructions — Spout

1. Lay out and cut View C.
2. Break in.
3. Make full folds for 4 mm grooved seam.
4. Shape the spout on a suitable stake.
5. Complete the grooved seam.
6. Flare 5 mm flange to suit body.
7. Place spout in position (outside body) and solder.
8. Solder grooved seams.
9. Dress all sharp edges.